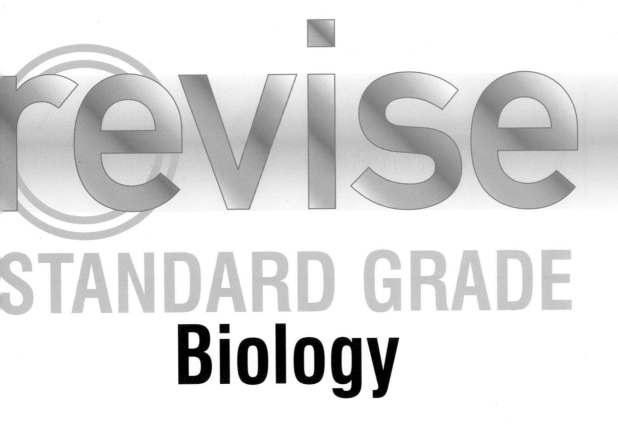

revise
STANDARD GRADE
Biology

David Applin

with Nicky Souter and Tony Buzan

Hodder & Stoughton

A MEMBER OF THE HODDER HEADLINE GROUP

Key to symbols

As you read through this book you will notice the following symbols. They will help you find your way around the book more quickly.

  shows a handy hint to help you remember something

 shows you a short list of key facts

 means remember!!!

 says Did you know this? – interesting points to note

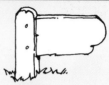 points you to other parts of the book where related topics are explained

 shows a sequence of linked processes

refers you from a diagram to a checklist of related points

shows content aimed at credit level

Acknowledgements

Copyright photographs have been used, with permission, from the Science Photo Library (pp. 89, 116, 117).

ISBN 0 340 77148 8

First published 2000
Impression number 10 9 8 7 6 5 4 3 2 1
Year 2006 2005 2004 2003 2002 2001 2000

Designed and produced by Gecko Ltd, Bicester, Oxon.
Printed in Spain for Hodder & Stoughton Educational, a division of Hodder Headline Plc,
338 Euston Road, London NW1 3BH, by Graphycems.

Project manager: Jo Kemp
Consultant: Nicky Souter
Mind Maps: Patrick Mayfield, Gareth Morris, Vanda North
Illustrations: Peter Bull, Simon Cooke, Chris Etheridge,
 Ian Law, Joe Little, Andrea Norton, Mike Parsons,
 John Plumb, Dave Poole, Chris Rothero, Anthony Warne
Cover design: Amanda Hawkes
Cover illustration: Paul Bateman

Contents

Revision
made easy

The four pages that follow contain a gold mine of information on how you can achieve success both at school and in your exams. Read them and apply the information, and you will be able to spend less, but more efficient, time studying, with better results. If you already have another Hodder & Stoughton revision guide, skim-read these pages to remind yourself about the exciting new techniques the books use, then move ahead to page 8.

This section gives you vital information on how to remember more *while* you are learning and how to remember more *after* you have finished studying. It explains

> **how to use special techniques to improve your memory**

> **how to use a revolutionary note-taking technique called Mind Maps that will double your memory and help you to write essays and answer exam questions**

> **how to read everything faster while at the same time improving your comprehension and concentration**

All this information is packed into the next four pages, so make sure you read them!

Your *amazing* memory

There are five important things you must know about your brain and memory to revolutionise your school life.

> **1 how your memory ('recall') works *while* you are learning**

> **2 how your memory works *after* you have finished learning**

> **3 how to use Mind Maps – a special technique for helping you with all aspects of your studies**

> **4 how to increase your reading speed**

> **5 how to zap your revision**

1 Recall during learning – the need for breaks

When you are studying, your memory can concentrate, understand and remember well for between 20 and 45 minutes at a time. Then it *needs* a break. If you carry on for longer than this without one, your memory starts to break down! If you study for hours non-stop, you will remember only a fraction of what you have been trying to learn, and you will have wasted valuable revision time.

So, ideally, *study for less than an hour*, then take a five- to ten-minute break. During the break listen to music, go for a walk, do some exercise, or just daydream. (Daydreaming is a necessary brain-power booster – geniuses do it regularly.) During the break your brain will be sorting out what it has been learning, and you will go back to your books with the new information safely stored and organised in your memory banks. We recommend breaks at regular intervals as you work through the revision guides. Make sure you take them!

2 Recall after learning – the waves of your memory

What do you think begins to happen to your memory straight *after* you have finished learning something? Does it immediately start forgetting? No! Your brain actually *increases* its power and carries on remembering. For a short time after your study session, your brain integrates the information, making a more complete picture of everything it has just learnt. Only then does the rapid decline in memory begin, and as much as 80 per cent of what you have learnt can be forgotten in a day.

However, if you catch the top of the wave of your memory, and briefly review (look back over) what you have been revising at the correct time, the memory is stamped in far more strongly, and stays at the crest of the wave for a much longer time. To maximise your brain's power to remember, take a few minutes and use a Mind Map to review what you have learnt at the end of a day. Then review it at the end of a week, again at the end of a month, and finally a week before the exams. That way you'll ride your memory wave all the way to your exam – and beyond!

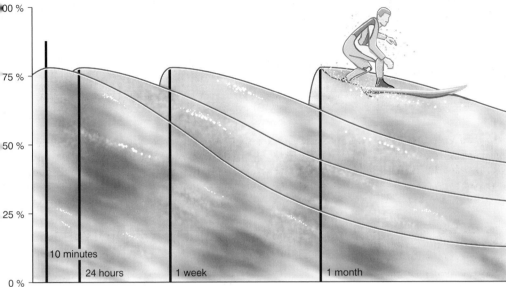

100 %

75 %

50 %

25 %

0 %

10 minutes

24 hours

1 week

1 month

review time

...azing as your memory is (think of ...rything you actually do have stored ...our brain at this moment) the ...ciples on which it operates are ...y simple: your brain will remember ... (a) has an image (a picture or a ...bol); (b) has that image fixed and ...can link that image to something else.

3 The Mind Map® – a picture of ...e way you think

...you *like* taking notes? More ...portantly, do you like having to go ...ck over and learn them before ...ms? Most students I know certainly ...not! And how do you take your ...es? Most people take notes on ...d paper, using blue or black ink. ...e result, visually, is *boring*! And what ...es your brain do when it is bored? It ...s off, tunes out, and goes to sleep! ...d a dash of colour, rhythm, ...gination, and the whole note-taking ...cess becomes much more fun, uses ...re of your brain's abilities, *and* ...roves your recall and understanding.

...Mind Map mirrors the way your brain ...rks. It can be used for note-taking ...n books or in class, for reviewing ...at you have just studied, for revising, ...d for essay planning for coursework ...d in exams. It uses all your memory's ...ural techniques to build up your ...idly growing 'memory muscle'.

You will find Mind Maps throughout this book. Study them, add some colour, personalise them, and then have a go at drawing your own – you'll remember them far better! Put them on your walls and in your files for a quick-and-easy review of the topic.

How to draw a Mind Map

1. Start in the middle of the page with the page turned sideways. This gives your brain the maximum room for its thoughts.

2. Always start by drawing a small picture or symbol. Why? Because a picture is worth a thousand words to your brain. And try to use at least three colours, as colour helps your memory even more.

3. Let your thoughts flow, and write or draw your ideas on coloured branching lines connected to your central image. These key symbols and words are the headings for your topic. The Mind Map at the top of the next page shows you how to start.

4. Then add facts and ideas by drawing more, smaller, branches on to the appropriate main branches, just like a tree.

5. Always print your word clearly on its line. Use only one word per line. The Mind Map at the foot of the

next page shows you how to do this.

6. To link ideas and thoughts on different branches, use arrows, colours, underlining, and boxes.

How to read a Mind Map

1. Begin in the centre, the focus of your topic.

2. The words/images attached to the centre are like chapter headings, read them next.

3. Always read out from the centre, in every direction (even on the left-hand side, where you will have to read from right to left, instead of the usual left to right).

Using Mind Maps

Mind Maps are a versatile tool – use them for taking notes in class or from books, for solving problems, for brainstorming with friends, and for reviewing and revising for exams – their uses are endless! You will find them invaluable for planning essays for coursework and exams. Number your main branches in the order in which you want to use them and off you go – the main headings for your essay are done and all your ideas are logically organised!

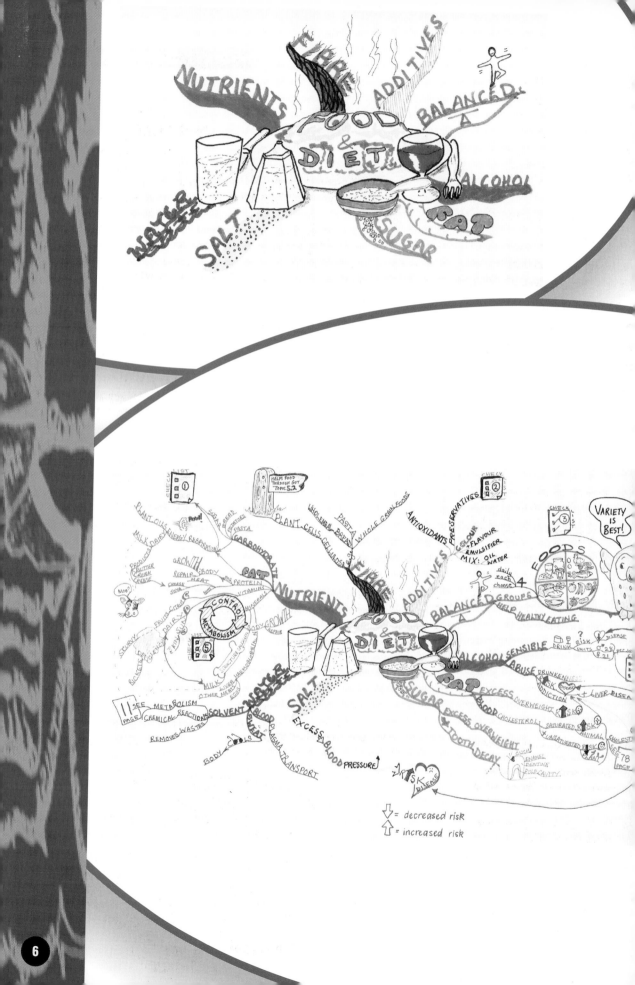

↓ = decreased risk

↑ = increased risk

4 Super speed reading

It seems incredible, but it's been proved – the faster you read, the more you understand and remember! So here are some tips to help you to practise reading faster – you'll cover the ground more quickly, remember more, *and* have more time for revision!

- First read the whole text (whether it's a lengthy book or an exam paper) very quickly, to give your brain an overall idea of what's ahead and get it working. (It's like sending out a scout to look at the territory you have to cover – it's much easier when you know what to expect!) Then read the text again for more detailed information.

- Have the text a reasonable distance away from your eyes. In this way your eye/brain system will be able to see more at a glance, and will naturally begin to read faster.

- Take in groups of words at a time. Rather than reading 'slowly and carefully' read faster, more enthusiastically. Your comprehension will rocket!

- Take in phrases rather than single words while you read.

- Use a guide. Your eyes are designed to follow movement, so a thin pencil underneath the lines you are reading, moved smoothly along, will 'pull' your eyes to faster speeds.

5 Helpful hints for exam revision

Start to revise at the beginning of the course. Cram at the start, not the end and avoid 'exam panic'!

Use Mind Maps throughout your course, and build a Master Mind Map for each subject – a giant Mind Map that summarises everything you know about the subject.

Use memory techniques such as mnemonics (verses or systems for remembering things like dates and events, or lists).

Get together with one or two friends to revise, compare Mind Maps, and discuss topics.

And finally...

- ⭐ *Have fun while you learn* – studies show that those people who enjoy what they are doing understand and remember it more, and generally do it better.

- ⭐ *Use your teachers* as resource centres. Ask them for help with specific topics and with more general advice on how you can improve your all-round performance.

- ⭐ *Personalise your copy of this book* by underlining and highlighting, by adding notes and pictures. Allow your brain to have a conversation with it!

Your brain is an amazing piece of equipment – learn to use it, and you, like thousands of students before you will be able to master General Level and Credit Level where appropriate. The more you understand and use your brain, the more it will repay you!

Before you begin

Revise Standard Grade Biology will help you prepare for tests and examinations. The book clearly and simply sets out what you have to **know** and **understand** for your success in standard grade biology. Examples of **problem solving** will help you build on knowledge and understanding and support the **practical abilities** you have developed in the laboratory.

The syllabus sets out *suggested learning activities*, which are a framework for the development of problem solving and practical abilities. *Learning outcomes* describes the performance expected in knowledge and understanding at general level and credit level.

Matching *Revise Standard Grade Biology* with the syllabus

The syllabus for standard grade biology consists of seven topics:

- The biosphere
- The world of plants
- Animal survival
- Investigating cells
- The body in action
- Inheritance
- Biotechnology

For each topic there are a number of sub-topics. Each sub-topic is made up of a set of numbered items. The grid below follows the pattern of the syllabus, matching a short summary of each numbered item with the relevant pages in this book. The letters G and Cr identify content at General Level and Credit Level where appropriate.

You can use the summaries as a checklist of your progress through the syllabus. Tick and date the boxes as you complete each item. Your confidence of success will grow as the list of ticks grows.

STANDARD GRADE BIOLOGY PLANNER	Level	Page	✓	Date
Topic 1: The biosphere				
Sub-topic (a) Investigating an ecosystem				
1 the main parts of an ecosystem	G	19–21		
2 techniques for sampling organisms	G	24–25		
sources of error	Cr	24–25		
3 using keys		17–18		
4/5 abiotic factors and their measurement	G	22–23		
6 abiotic influences on the distribution of organisms	G	22–23		
abiotic mechanisms affecting distribution	Cr	20, 22		

STANDARD GRADE BIOLOGY PLANNER	Level	Page	✓	Date
Sub-topic (b) How it works				
7 the meaning of ecological terms	G	19–21		
8 producers and consumers	G	27		
9 food chains and food webs	G	26–27		
removing a species from a food web	Cr	28		
0 the flow of energy through a food web	G	27, 29		
pyramid of numbers/biomass	Cr	28–29		
1 the size of a population	G	30		
describing population growth	Cr	30–31		
2 limiting factors on population growth	G	30		
explaining population growth	Cr	30–31		
3 competition and its effects	G	26		
4 the importance of nutrient cycles	G	21, 125		
the nitrogen cycle	Cr	21		
Sub-topic (c) Control and management				
5 pollution affects air, water and land	G	31–33		
adverse effects of fossil fuels and nuclear power as energy sources	Cr	32–33		
the main sources of pollution are domestic, agricultural and industrial	G	31–33		
6 controlling pollution	G	32		
7 organic waste is food for microorganisms	G	33		
effects of organic waste on different organisms	Cr	33–34		
indicator species	Cr	34		
8 improving the management of natural resources	G	35–36		
agriculture and forestry	Cr	36–38		
problems of poor management of natural resources	G	35–36		
Topic 2: The world of plants				
Sub-topic (a) Introducing plants				
1 the variety of plants and its advantages	G	40–41		
the consequences of reducing plant variety	Cr	43		
2 uses of plants	G	42, 44–45		
producing and refining plant products	Cr	42		
using plant products	Cr	42		
Sub-topic (b) Growing plants				
3 parts of a seed	G	47–48		
4 factors affecting germination	G	48–49		
percentage germination	Cr	48		
5 parts of a flower	G	43		
the structure of wind pollinated and insect pollinated flowers	Cr	47		
methods of pollination	G	46–47		
growth of a pollen tube	Cr	46		
fertilisation and fruit formation	G	49		
6 dispersal of fruits and seeds	Cr	49		

STANDARD GRADE BIOLOGY PLANNER	Level	Page	✓	Date
7 cuttings and grafting	G	51		
advantages of artificial propagation	Cr	51		
What is a clone?	Cr	51, 88		
8 asexual reproduction in flowering plants	G	49–50		
advantages of sexual and asexual reproduction to flowering plants	Cr	51–52		
Sub-topic (c) Making food				
9 transport in plants	G	56–58		
the structure and function of xylem and phloem	Cr	56–57		
the movement of water and food	G	56–57		
10 air enters the leaf through the stomata	G	56–57		
the leaf is adapted for gaseous exchange	Cr	55		
11 water vapour is lost from the leaf through the stomata	G	57–58		
12 making food	G	52		
the uses of sugar in plants	Cr	52		
13 chlorophyll traps light energy	G	52, 54		
photosynthesis	G	52, 54–55		
limiting factors for photosynthesis	Cr	53		
Topic 3: Animal survival				
Sub-topic (a) The need for food				
1 Why do animals need food?	G	6, 60		
carbohydrates, proteins and fats	Cr	61–62		
2 digestion allows food to be absorbed into the blood stream	G	63		
digestion converts insoluble food substances into soluble food substances	Cr	63, 66		
3 the types and role of teeth in a herbivore, carnivore and an omnivore	G	67–68		
4 the parts of the alimentary canal (digestive system)	G	64–65		
digestive juices	Cr	63, 66		
peristalsis and the stomach	Cr	64		
5 digestive enzymes	G	66		
the activities of an amylase, protease and lipase	Cr	66		
6 the structure and function of the small intestine	G	63–65		
villi and the absorption and transport of food	Cr	64, 66		
7 the large intestine absorbs water and eliminates waste	G	65		
Sub-topic (b) Reproduction				
8 the features of sperm and eggs	G	68–69		
9 external fertilisation and internal fertilisation	G	69		
the importance of internal fertilisation to land-living animals	Cr	69		
the process of fertilisation	G	69, 72		
10 sperm cells are produced in testes	G	71		
eggs are produced in ovaries	G	71		
fertilisation occurs in an oviduct	G	71		
11 fish eggs	G	69		
survival of young	Cr	69–70		

STANDARD GRADE BIOLOGY PLANNER	Level	Page	✓	Date
Sub-topic (d) Investigating enzymes				
11 Why do cells need enzymes?	G	88		
What is a specificity?	Cr	88		
What is a catalyst?	G	88		
12/13 enzymes catalyse the breakdown and synthesis of substances	G	91		
14 enzymes are proteins	G	88		
the effect of temperature on enzyme activity	G	90–91		
What does 'optimum' mean?	Cr	90		
the effect of pH on the activity of pepsin and catalase	G	90		
Sub-topic (e) Investigating aerobic respiration				
15 cells use energy	G	87, 91–92		
16 the energy content of fats and oils is greater than carbohydrates or proteins	Cr	92		
17 oxygen is needed to release energy from food during aerobic respiration	G	87, 91		
18 respiration gives off carbon dioxide	G	87, 91		
19 heat is released during respiration	G	91		
Why do cells need energy?	Cr	92		
Topic 5: The body in action				
Sub-topic (a) Movement				
1/2 the skeleton is a framework for support and muscle attachment	G	94		
the skeleton protects vital organs	G	95		
3 joints allow movement	G	95		
the functions of ligaments and cartilage	G	94, 96		
the structure and function of a synovial joint	Cr	95		
4 bone consists of flexible fibres and hard minerals	G	95		
bone is formed by cells	Cr	95		
5 tendons attach muscles to bones	G	94		
tendons are inelastic	Cr	96		
6 muscle contraction brings about movement	G	95		
opposing muscles	Cr	96		
Sub-topic (b) The need for energy				
7 balancing energy input and output	G	96		
8 oxygen is absorbed and carbon dioxide released during breathing	G	98		
9 the internal structure of the lungs	G	97		
the mechanism of breathing	Cr	99		
the function of cilia, cartilage and mucus in the trachea and bronchi	Cr	97–98		
gas exchange	Cr	97		
Why are the lungs efficient gas exchange structures?	Cr	98		
10 the heart has four chambers	G	101		
the flow of blood through the heart and its associated blood vessels	G	101		
the heart valves	G	101		

STANDARD GRADE BIOLOGY PLANNER	Level	Page	✓	Date
the walls of the ventricles	G	102		
the coronary arteries supply the heart with blood	G	102		
1 arteries transport blood from the heart, veins transport blood to the heart and capillaries join arteries with veins	G	99, 101		
the pulse	G	101, 109		
2 the function of red blood cells and plasma	G	86, 100		
the function of haemoglobin	Cr	102		
exchanges between cells and capillaries	G	102–103		
the features of a capillary network	Cr	102–103		
Sub-topic (c) Co-ordination				
3 the components of co-ordination		103		
4 judging distance	G	108		
binocular vision	Cr	108		
5 the structure of the mammalian eye	G	104, 107		
6 judging the direction of sound	G	104		
7 the structure of the ear	G	104, 107		
the semi-circular canals	Cr	107		
8 the structure of the nervous system	G	105, 106		
9 the flow of information: senses → central nervous system → muscles	G	103		
the reflex arc	Cr	103, 105		
the central nervous system processes information	Cr	105		
20 the functions of the cerebrum	Cr	105		
cerebellum and medulla				
Sub-topic (d) Changing levels of performance				
21 muscle fatigue	G	109		
muscle fatigue is the result of anaerobic respiration	Cr	109		
22 exercise increases the pulse rate and breathing rate	G	109		
23 an athlete v an untrained person	G	110		
recovery time	G	110		
the effects of training	Cr	109		
recovery time is used as a measure of fitness	G	110		
the relationship between the effects of training and recovery time	Cr	110		
Topic 6: Inheritance				
Sub-topic (a) Variation				
1 What is a species?	G	112		
2 variation occurs within a species	G	112		
3 examples of continuous and discontinuous variation	G	113		
What is meant by continuous and discontinuous variation?	Cr	113		
Sub-topic (b) What is inheritance?				
4 parental genes determine characteristics in offspring	G	115		
5 examples of phenotypes	G	115–116		
identifying true-breeding, dominant and recessive characteristics	G	114–115		

STANDARD GRADE BIOLOGY PLANNER	Level	Page	✓	Date
6 identifying P, F$_1$ and F$_2$ generations	G	114		
the parents in monohybrid crosses are usually true-breeding	Cr	114		
the phenotype of the F$_1$ generation are the same	G	114		
predicting the proportions of phenotypes in the F$_2$ generation	Cr	114		
7 each body cell has 2 matching sets of chromosomes	G	115		
sex cells are called gametes	G	115		
chromosomes are reduced to a single set during gamete formation	G	88, 115		
each sex cell carries one set of chromosomes	G	115		
fertilisation restores a double set of chromosomes	G	88, 115		
8 genes are parts of chromosomes	G	115		
a characteristic is controlled by two forms of a gene	G	114–115		
different forms of a gene are called alleles	Cr	115		
each parent contributes one of the two forms of a gene	G	88, 115		
each gamete carries one of the two forms of a gene	G	88, 114–115		
What does 'genotype' mean?	G	115		
monohybrid crosses expressed in terms of genotypes	Cr	115		
Why are there differences between observed and predicted results in monohybrid crosses?	Cr	115–116		
9 the X and Y chromosomes determine the sex of a child	G	116		
each male gamete may have an X or a Y chromosome; each female gamete has an X chromosome	G	116		
how the sex of a child is determined	G	116		
10/11 selective breeding improves characteristics	G	116–117		
examples of the improvement of a characteristic through selective breeding	Cr	116–117		
12 an example of a chromosome mutation in humans	G	117		
an example of a chromosome mutation in a plant or animal which is to human advantage	Cr	118		
amniocentesis can detect chromosome characteristics before birth	G	118		
factors influencing the rate of mutation	Cr	112		
Topic 7: Biotechnology				
Sub-topic (a) Living factors				
1 the raising of dough and the manufacture of beer and wine depend on the activities of yeast	G	120–123		
2 yeast is a single-celled fungus which uses sugar as food	G	121		
3 the process of fermentation	G	120		
comparing anaerobic respiration with aerobic respiration	Cr	91–92,109,120		
4 the best growing conditions for yeast	Cr	121		
What is batch processing?	Cr	121		
5 malting barley	Cr	42, 121		
6 making cheese and yoghurt depends on the activities of bacteria	G	122		
7 the souring of milk is a fermentation process	G	121		
milk sours because bacteria ferment lactose	Cr	121		

Introducing biology

How much do you already know?
Work out your score on page 134.

Test yourself

1 What would happen to the ground temperature if Earth were
   a) nearer to the Sun
   b) further from the Sun? [2]

2 a) List the processes which tell you that something is living. $[7 \times \frac{1}{2}]$
   b) Put a tick (✓) next to the processes which you think apply to animals. $[7 \times \frac{1}{2}]$
   c) Put a cross (✗) next to the processes which you think apply to plants. $[6 \times \frac{1}{2}]$
   d) Do plants and animals have the same characteristics? If not, how are they different? [1]

3 a) What is a biological key used for? [1]
   b) Why are features like exact colour, size and mass not suitable for including in a biological key? [3]

A Living on Earth

preview

At the end of this section you will:
- **understand why Earth is a suitable place fo living things (organisms)**
- **know that soil, air and water are the physic environments in which organisms live.**

Why the Earth can support life

Earth is a planet in orbit round a star we call th Sun. It is the only planet we know of that suppo life.

★ Earth is close enough to the Sun for its surface temperature to be in the range in which life can exist. The temperature at the Earth's surface vari between –70 and 55°C.

★ Earth is massive enough to have sufficient gravit to hold down an atmosphere of different gases essential for living organisms.

★ The layer of ozo which surrounds Earth reduces th amount of ultraviolet light fr the Sun reaching the planet's surface. Too mu ultraviolet light destroys living things.

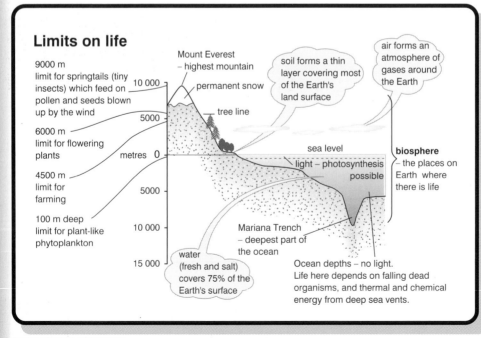

Limits on life

9000 m limit for springtails (tiny insects) which feed on pollen and seeds blown up by the wind

Mount Everest – highest mountain

permanent snow

tree line

soil forms a thin layer covering most of the Earth's land surface

air forms an atmosphere of gases around the Earth

10 000

5000

6000 m limit for flowering plants

metres 0

sea level

light – photosynthesis possible

biosphere – the places on Earth where there is life

4500 m limit for farming

5000

100 m deep limit for plant-like phytoplankton

10 000

Mariana Trench – deepest part of the ocean

15 000

water (fresh and salt) covers 75% of the Earth's surface

Ocean depths – no light. Life here depends on falling dead organisms, and thermal and chemical energy from deep sea vents.

Earth's physical environment

e diagram on the opposite page shows that soil,
and water form Earth's environment:

Soil is formed when the weather, the roots of plants
and the different activities of animals break down
rocks into small particles.

Air consists of: 78% nitrogen; 21% oxygen; 0.03%
carbon dioxide; and less than 1% water vapour,
argon, xenon and other gases.

Water fills the seas, oceans, rivers and lakes. About
2% of the Earth's water is locked up as ice, in the
soil, in the bodies of living things or is vapour in the
atmosphere.

Characteristics of life

preview

At the end of this section you will know the characteristics of life:
- **Movement**
- **Respiration**
- **Sensitivity**
- **Growth**
- **Reproduction**
- **Excretion**
- **Nutrition.**

andy hint

Hints & Tips

e mnemonic **Mrs Gren** will help you
member the characteristics of living things.

lore about MRS GREN

e characteristics of life are the features
at are common to all living things:

Movement: animals are able to move from place to
place because of the action of **muscles** which pull
on the **skeleton**. Plants do not usually move from
place to place; they move mainly by **growing**.

Respiration occurs in cells, and releases energy
from food for life's activities. **Aerobic** respiration
uses oxygen to release energy from food.
Anaerobic respiration releases energy from food
without using oxygen.

Sensitivity allows living things to detect changes in
their surroundings and respond to them.

★ **Growth** leads to an increase in size. **Development**
occurs as young change and become adult in
appearance.

★ **Reproduction** produces new individuals.

★ **Excretion** removes the waste substances produced
by the chemical reactions (called **metabolism**)
taking place in cells.

★ **Nutrition** makes food (by the process of
photosynthesis) or takes in food for use in the body.

Remember
- **Respiration** releases energy from food.
- **Gaseous exchange** takes in oxygen for
  respiration and removes carbon dioxide
  produced by respiration.
- **Excretion** removes wastes produced by
  metabolism.
- **Defecation** (or egestion) removes the
  undigested remains of food.

C Using keys

preview

At the end of this section you will:
- know that a key is a set of clues that help
  identify a particular organism or group of
  organisms
- understand how to use a dichotomous key
- know that a dichotomous key can be written
  in different ways.

What is a key?

A **key** is a means of identifying an unfamiliar
organism from a selection of specimens. A key
consists of a set of descriptions. Each description
is a clue that helps in the identification. A set of
clues makes the key.

The easiest type of key to use is called a
dichotomous key. 'Dichotomous' means branching
into two. Each time the key branches, you have to
choose between alternative statements. The
alternative statements may be presented
diagrammatically as a chart, or written in pairs or

couplets. For example, a key to amphibians would begin:

		yes	**no**
1	The animal has a tail.	**newts**	go to **2**
2	The animal has no tail.	**frogs and toads**	go to **3**

and so on …

By comparing the pairs of statements with the organisms, you will eventually find one that fits. This identifies the organism. A key is therefore the route to a name. Different keys are used to name different living things.

When making a key, it is important to choose features that are characteristic of the type of organism rather than of the individual itself. Fo example, shape or proportions and patterns of colour are fairly constant in a type of organism and are therefore useful clues in a key. Size and shades of colour vary from individual to individual and are of limited use.

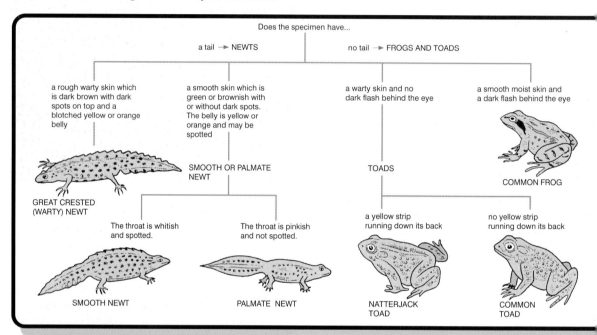

A key to amphibians

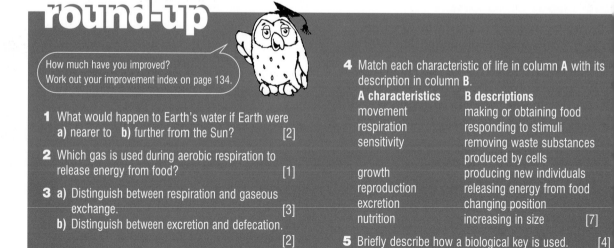

round-up

How much have you improved?
Work out your improvement index on page 134.

1 What would happen to Earth's water if Earth were
a) nearer to   **b)** further from the Sun?   [2]

2 Which gas is used during aerobic respiration to release energy from food?   [1]

3 a) Distinguish between respiration and gaseous exchange.   [3]
   b) Distinguish between excretion and defecation.   [2]

4 Match each characteristic of life in column **A** with its description in column **B**.

A characteristics	B descriptions
movement	making or obtaining food
respiration	responding to stimuli
sensitivity	removing waste substances produced by cells
growth	producing new individuals
reproduction	releasing energy from food
excretion	changing position
nutrition	increasing in size [7]

5 Briefly describe how a biological key is used.   [4]

The biosphere

How much do you already know? Work out your score on page 134.

Test yourself

Match each term in column **A** with the correct description in column **B**.

A terms	B descriptions
biosphere	the place where a group of organisms lives
community	all the ecosystems of the world
habitat	a group of individuals of the same species
population	all the organisms that live in a particular ecosystem [4]

a) Why is a food web a more accurate description of feeding in a community than a food chain? [2]

b) Why do food chains and food webs always begin with plants? [4]

Why is the pyramid of biomass usually a better description of a community than the pyramid of numbers? [2]

The table below shows the change in numbers of sheep for 105 years since their introduction into South Australia.

Year	Number of sheep (millions)
1830	0.1
1840	0.2
1850	1.0
1860	3.0
1870	5.0
1880	6.5
1890	6.9
1900	5.2
1910	6.1
1920	7.4
1930	6.0
1935	8.0

a) Plot the data on graph paper. [8]

b) Between which years was population growth proportional to the numbers of sheep? [2]

c) In which years did limiting factors stop the growth in numbers of sheep? [2]

d) What was the overall effect of limiting factors on the numbers of sheep between 1890 and 1935? [1]

Summarise the effects of organic waste on the abiotic and biotic parts of a river ecosystem. [5]

• •

1.1 Introducing ecology

preview

At the end of this section you will understand that:

● an ecosystem is a self-contained part of the biosphere, such as a pond or an oak wood

● the community consists of the organisms that live in a particular ecosystem

● the habitat is the place where organisms live

● a population is a group of organisms of the same species living in a particular place at the same time

● decomposition releases mineral nutrients from dead organic matter into the environment for recycling.

Some ecological terms

Ecology involves studying the relationships between organisms and between organisms and the environment.

The diagram on page 16 shows how all the places on Earth where there is life form the **biosphere**. Each organism is suited (adapted) to the place where it lives. This place consists of

• a non-living (**abiotic**) **environment** of air/soil/water

• a living (**biotic**) community of plants, animals, fungi and microorganisms.

Environment and the community together form an **ecosystem**, which is a more or less self-contained part of the biosphere. 'Self-contained' means that each ecosystem has its own characteristic organisms not usually found in other ecosystems. The diagram on the following page shows the different components of an oak wood ecosystem.

BIOTIC

key
1 oak tree
2 hazel
3 holly
4 bluebell
5 wood anemone
6 primrose
7 moss on tree trunk
8 pigeons, rooks living in canopy
9 bluetits, woodpeckers living further down tree
10 great tits, warblers living in shrubs
11 wrens, blackbirds living on ground
12 toadstools on rotting log
13 woodlice in detritus
14 earthworm pulling leaf into burrow
↙ falling leaves

HABITATS
canopy layer

ABIOTIC
There may be up to 90% less light insid
than outside when the canopy is fully de

THE ECOSYSTEM

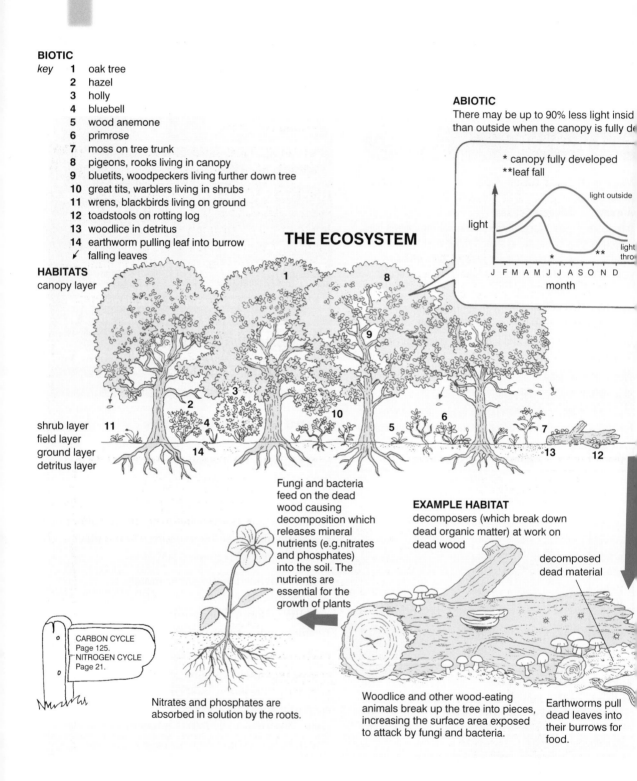

* canopy fully developed
**leaf fall

light

light outside

light
thro

J F M A M J J A S O N D
month

shrub layer
field layer
ground layer
detritus layer

Fungi and bacteria feed on the dead wood causing decomposition which releases mineral nutrients (e.g.nitrates and phosphates) into the soil. The nutrients are essential for the growth of plants

EXAMPLE HABITAT
decomposers (which break down dead organic matter) at work on dead wood

decomposed dead material

CARBON CYCLE
Page 125.
NITROGEN CYCLE
Page 21.

Nitrates and phosphates are absorbed in solution by the roots.

Woodlice and other wood-eating animals break up the tree into pieces, increasing the surface area exposed to attack by fungi and bacteria.

Earthworms pull dead leaves into their burrows for food.

The components of an oak wood ecosystem

The flow chart below shows the hierarchy of ecological terms.

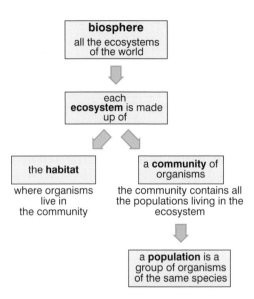

biosphere
all the ecosystems of the world

each **ecosystem** is made up of

the **habitat**
where organisms live in the community

a **community** of organisms
the community contains all the populations living in the ecosystem

a **population** is a group of organisms of the same species

Notice in the diagram of an oak wood ecosystem that

★ decomposition releases mineral nutrients into the soil

★ the mineral nutrients are **absorbed** in solution by the roots of plants

as a result plants obtain the mineral nutrients essential for growth.

Nitrates are nutrients used by living things to make **protein**. The element **nitrogen** is present in nitrates (e.g. calcium nitrate). It is recycled from living things to the air, soil and water and back again. Other elements essential for life are also recycled between organisms and the environment. The diagram gives you the idea.

PROTEINS Page 62.
CARBON CYCLE Page 125.

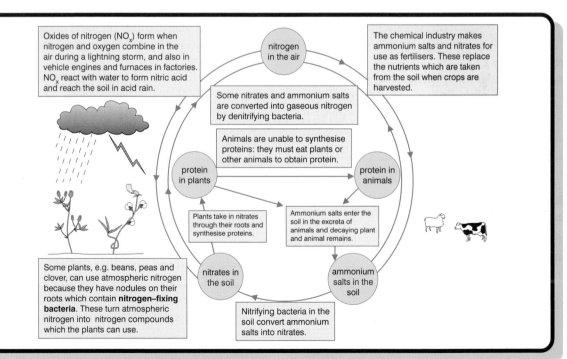

Oxides of nitrogen (NO_x) form when nitrogen and oxygen combine in the air during a lightning storm, and also in vehicle engines and furnaces in factories. NO_x react with water to form nitric acid and reach the soil in acid rain.

nitrogen in the air

The chemical industry makes ammonium salts and nitrates for use as fertilisers. These replace the nutrients which are taken from the soil when crops are harvested.

Some nitrates and ammonium salts are converted into gaseous nitrogen by denitrifying bacteria.

Animals are unable to synthesise proteins: they must eat plants or other animals to obtain protein.

protein in plants

protein in animals

Plants take in nitrates through their roots and synthesise proteins.

Ammonium salts enter the soil in the excreta of animals and decaying plant and animal remains.

Some plants, e.g. beans, peas and clover, can use atmospheric nitrogen because they have nodules on their roots which contain **nitrogen–fixing bacteria**. These turn atmospheric nitrogen into nitrogen compounds which the plants can use.

nitrates in the soil

ammonium salts in the soil

Nitrifying bacteria in the soil convert ammonium salts into nitrates.

Nitrogen circulates from air to soil to living things and back again in the nitrogen cycle

1.2 Investigating an ecosystem

preview

At the end of this section you will:

- be able to describe techniques for sampling organisms
- be able to identify and measure different abiotic factors
- know the effects that different factors have on the distribution of organisms.

Look back at Section 1.1 (pages 19–21) to help you remember the different parts of an ecosystem.

Most ecosystems are too big for you to study all of the living things present. You can only study different parts and assume that the parts are representative of the whole ecosystem. The parts studied are called **samples**. The concept map on pages 24 and 25 shows different ways of taking samples of organisms in the different habitats of a woodland ecosystem.

Remember

Organisms are the living (**biotic**) community of the ecosystem. Different methods are used to measure the non-living (**abiotic**) parts of the ecosystem.

In a woodland ecosystem light intensity, temperature of the soil and air, and soil moisture are important abiotic factors.

Sampling abiotic factors in a woodland ecosystem

★ *Errors in comparing the abiotic factors in different parts of the ecosystem:*
  - samples are taken at different times of the day (or night)
  - samples are taken in different conditions of weather.

★ *Errors in using equipment:*
  - thermometers and/or meters (light/temperature/pH) are not given long enough to provide a stable reading

  - thermometers and probes (light/temperature/pH) are not pushed into the soil to the same depth each time in a series of measurements
  - shadow is cast over a light meter.

★ *Reducing sample errors:*

  - when comparing abiotic factors in different parts of the ecosystem, take measurements under similar conditions (e.g. weather conditions; time of day or night)
  - allow thermometers and probes (light/temperature/pH) enough time to stabilise before taking measurements
  - take a number of readings for a particular abiotic factor in each part of the ecosystem sampled. Work out the average for each set of readings
  - establish a *consistent* pattern of working so that measurements can be relied upon to give a fair comparison of the abiotic factors in different parts of the ecosystem.

Distribution of organisms

The distribution of organisms (how many living things are found in different parts of the environment) is affected by abiotic factors.

Light

The intensity of light is an important abiotic influence on the distribution of organisms inside a wood (see page 20).

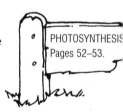
PHOTOSYNTHESIS Pages 52–53.

It affects the rate of photosynthesis and therefore the amount of plant growth under the canopy layer.

★ Plants of the field layer (see page 20) grow and flower in the spring. The plants take advantage of the increasing intensity of sunlight for photosynthesis. The sunlight reaches the woodland floor through branches bare of leaves.

As a result, more food and shelter are available for animals.

ain and temperature

th rain and warmth communities flourish. The gram shows the effect of seasonal rainfall on e distribution of plants in West Africa.

eather stations are located
the places named on the map.
nanges in the type of vegetation
e shown along a transect
ee page 24) marked A – B

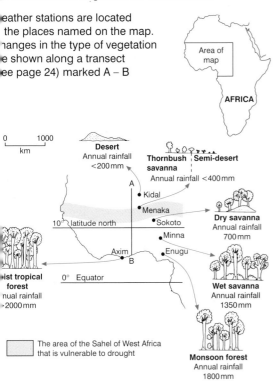

AFRICA

Area of map

(see page 24)

e factor that affects the distribution of plants
gion-by-region across the world is the different
nperature zones north and south of the equator.
e map shows the pattern.

A second factor is height above sea level. Compare the map with the distribution of plants up mountains located near to the equator, as shown in the diagram below.

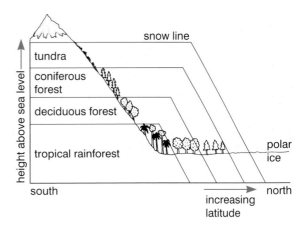

Remember that for every 100 metres climbing the mountainside, the temperature drops about 0.5°C. This is the same difference as the drop in temperature over a 1° increase in latitude north or south of the equator. This explains why living things normally found in colder regions may reach the equator by travelling the mountain ranges that run from north to south.

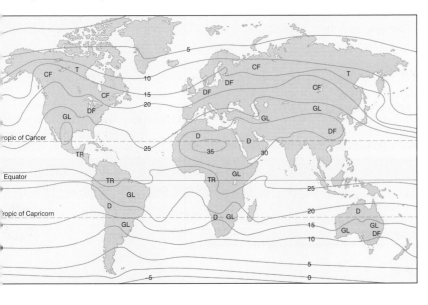

Key to major vegetation areas

TR Tropical rain forest

GL Grassland

D   Desert

DF Deciduous forest
(broad-leaved trees
eg oak)

CF Coniferous forest
(evergreens with
needle-like leaves eg fir)

T   Tundra (area too cold for
trees. Dominant plants are
lichens and mosses)

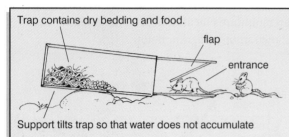

Trap contains dry bedding and food.

flap

entrance

Support tilts trap so that water does not accumulate

Longworth (small mammal) trap – a box baited with food and supplied with dry bedding and used to trap small mammals (e.g. voles, mice).
- Identify the animals caught.
- Release the animals from the trap.

Reducing sampling error:
- set the trap so that the animals can enter easily
- traps should be inspected regularly to reduce stress to the captive animals.

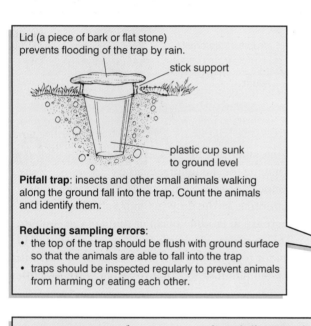

Lid (a piece of bark or flat stone) prevents flooding of the trap by rain.

stick support

plastic cup sunk to ground level

Pitfall trap: insects and other small animals walking along the ground fall into the trap. Count the animals and identify them.

Reducing sampling errors:
- the top of the trap should be flush with ground surface so that the animals are able to fall into the trap
- traps should be inspected regularly to prevent animals from harming or eating each other.

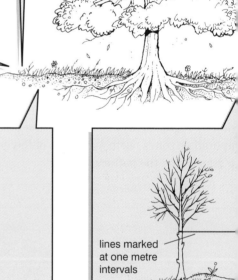

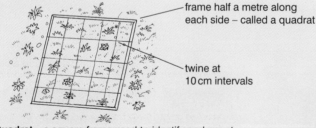

frame half a metre along each side – called a quadrat

twine at 10 cm intervals

lines marked at one metre intervals

Quadrat – a square frame used to identify and count the number of plants or animals in a *known* area.
- Throw the quadrat at random in the study area.
- Count the plants or animals and identify them.
- Calculate abundance as the number of squares that the plant or animal occupies.

Reducing sampling errors:
- take sufficient number of samples
- avoid choosing 'good looking' areas to sample.
  Remember, sampling must be *random*
- only count individual plants or animals that lie at least half inside each square of the quadrat, to estimate the abundance.

Sampling abiotic factors in a woodland ecosystem

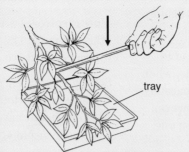

Tree beating: a stick is used the tap the branches of shrubs or tress, knocking small animals living there on to a tray or white cloth.
- Do not damage the shrub or tree.
- Count the animals and identify them.

tray

Reducing sampling errors:
- take a sufficient number of samples
- take samples following a *consistent* pattern (e.g. tap branches the same number of times for each sample)
- do not sample the same branch more than once.

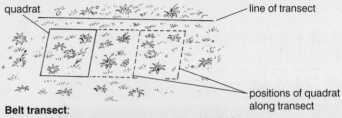

quadrat .... line of transect

.... positions of quadrat along transect

Belt transect:
– uses quadrat and line transect together. Useful for measuring changes in vegetation between two points.
- Lay down tape or rope along the line of the transect.
- Use a quadrat to record the plant species at intervals along the transect.
- Estimate abundance of each species.

Reducing sampling errors:
- the position of the quadrat along the line of transect should follow a *consistent* pattern.

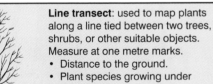

Line transect: used to map plants along a line tied between two trees, shrubs, or other suitable objects. Measure at one metre marks.
- Distance to the ground.
- Plant species growing under the mark.
- Height of plant(s).

Reducing sampling errors: only record the plants growing directly under each of the one metre marks along the line of transect.

Sweep net:
a strong cotton bag attached to a handle and swept through tall grass and other vegetation. 'Swishing' the net through the grass knocks small animals (e.g. insects, spiders) into the bag.
- Count the animals and identify them.

Reducing sampling errors:
- take a sufficient number of samples
- take samples following a *consistent* pattern (e.g. walk ten paces while 'swishing', taking one 'swish' for each step)
- do not sample the same area of vegetation more than once

Competition between organisms also affects their distribution. In nature competitors are rivals for factors that are in limited supply. Factors include resources like water, light, space, food or mates.

Competition for the same resource may lead to one species replacing another. For example, when two species of clover plant were grown separately in different containers, they grew well. However, when grown together in the same container, one species eventually replaced the other. The reason was that the successful species grew slightly taller than its competitor and overshadowed it. *Why do you think the unsuccessful species died out?*

1.3 Food chains and webs

preview

At the end of this section you will:
- **know the meaning of the terms producer, herbivore, carnivore and omnivore**
- **understand that energy is transferred along food chains**
- **be able to interpret diagrams of food chains and food webs.**

Feeding relationships

Finding out feeding relationships is one way of describing how a community works. Looking at

- animals' teeth or mouthparts
- what is in the intestine
- animals feeding
- what food an animal likes best

helps find out what animals feed on.

The working community

Animals fall into three categories according to what they eat.

★ **Herbivores** eat plants.

★ **Carnivores** eat meat.

★ **Omnivores** eat both plants and meat. (Most hum beings are omnivorous.)

Most carnivores are **predators** – they catch and e other animals. The animals caught are called their **prey**, and are often herbivores.

Scavengers are carnivores that feed on the remains of prey left by predators, or on the bodi of animals that have died for other reasons such as disease or old age.

A **food chain** shows the links between plants, pre predators and scavengers. Some examples of foc chains are shown opposite.

Notice in each example that

- the arrows represent the transfer of food energy between different organisms
- the arrows show the direction of energy flow
- the number of links in a food chain is usually four or less.

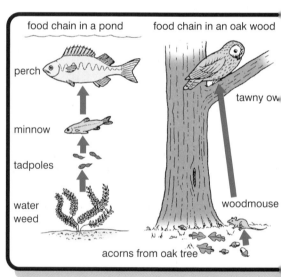

food chain in a pond   food chain in an oak wood

perch

minnow

tadpoles

water weed

tawny ow

woodmouse

acorns from oak tree

Food chains

A **food web** is usually a more accurate descriptio of feeding relationships in a community because most animals eat more than one type of plant or other animal. Some examples of food webs are shown opposite.

Notice in each example that

- several food chains link up to form a food we
- plants produce food by photosynthesis
- different animals eat the same type of food.

Producers

Plants, algae and some bacteria (see page 28) are
called **producers** because they can use sunlight to produce food by photosynthesis. This is why food chains and food webs always begin with plants (or algae or photosynthetic bacteria). Animals use this food when they eat plants. Even when they eat other animals, predators depend on plant food indirectly since somewhere along the line the prey has been a plant eater. Because they eat food, animals are called **consumers**.

Energy flow

Sunlight underpins life on earth. It is the ultimate source of energy for food chains and food webs. Without sunlight, and the photosynthesis which depends on it, communities would cease to exist.

Through photosynthesis plants convert light energy into the chemical energy of food. A food chain represents one pathway of food energy through the community of an ecosystem. A food web represents many pathways. The diagram below shows the idea.

PHOTOSYNTHESIS
Pages 52–54.

food web in a pond

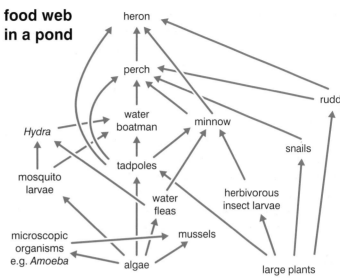

food web in an oak wood

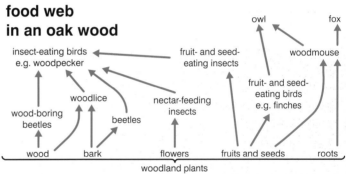

Food webs

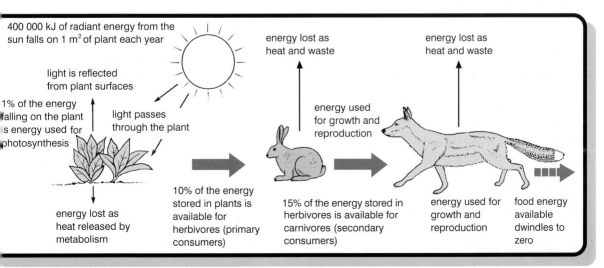

flow of energy through a food chain

At each link in the food chain, energy is lost as heat and in the waste products of each living thing. *What would happen to the pond's remaining organisms if the tadpoles were removed?*

1.4 Ecological pyramids

preview

At the end of this section you will:
- **know how to build a pyramid of numbers**
- **be able to interpret pyramids of numbers and pyramids of biomass**
- **understand that the pyramid of energy gives the best picture of the relationships between producers and consumers.**

How many?

Food chains and food webs describe the feeding relationships within a community. However, they do not tell us about the numbers of individuals involved. Many plants support a limited number of herbivores which in turn support fewer carnivores.

The idea is represented as different types of **ecological pyramid**. Organisms which have similar types of food are grouped together into **feeding levels**.

★ **Producers** (P) (plants/algae/some bacteria) occupy the feeding level that forms the base of the pyramid.

★ **Consumers** occupy the other feeding levels:
- herbivores (H) feed on plants
- primary carnivores (C_1) feed on herbivores
- secondary carnivores (C_2) feed on primary carnivores.

PRODUCERS
CONSUMERS
Page 27

Pyramids of numbers

Pyramids of numbers show the number of organisms in each feeding level. The table shows the numbers of insects and spiders collected from grass with a sweep net.

sample	number of insects	number of spiders
1	135	5
2	150	10
3	110	10
4	115	5
5	120	15
	630	45
	average $= \dfrac{630}{5}$ $= 126$ insects in each sample	average $= \dfrac{45}{5}$ $= 9$ spiders in each sample

Samples of insects and spiders collected with a sweep net. Each sample was collected with 10 sweeps while walking around an area of 10 m².

Data like that in the table can be used to plot a pyramid of numbers, as shown below. Half the number of organisms in each feeding level is plotted on one side of the vertical line, the other half is plotted on the other side of the vertical line.

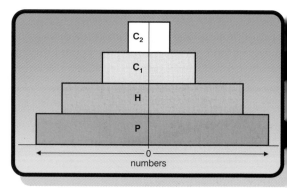

How to plot a pyramid of numbers

In grassland, the producers (grasses) and consumers (mainly insects and spiders) are small and numerous. A lot of plants support many herbivores (mostly insects) which in turn support fewer carnivores (mostly spiders). Plotting the number of organisms in each feeding level of the grassland community gives an upright pyramid that tapers to a point.

oblems with numbers ...

ample 1: the pyramid of numbers for a odland (see below) has a point at the bottom well as the top. This is because relatively few oducers (trees) support a large number of rbivores and carnivores. You might think that odland consumers are in danger of starvation! wever, each tree is large and can meet the d needs of many different organisms.

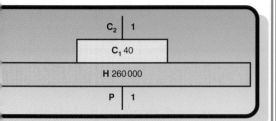

mid of numbers for a woodland community

ample 2: pyramids of numbers including rasites appear top-heavy, as shown below. Many rasites feed on fewer primary carnivores.

both these examples, number pyramids are an accurate description of the feeding ationships in the different communities. y? Because the pyramid of numbers does not e into account differences in size of the ferent producers and consumers.

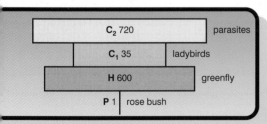

mid of numbers including parasites of the primary carnivores birds)

and the solution

yramid of biomass** allows for differences in the es of organisms, because the pyramid shows the *ss of organic material* in each feeding level.

A representative sample of the organisms at each feeding level is weighed.
The mass is then multiplied by the estimated number of organisms in the community.

practice, dry mass is used because fresh mass ries greatly as organisms contain different

amounts of water. The sample is dried out in an oven at 110°C until there is no further change in mass.

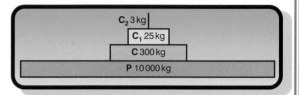

Pyramid of biomass for a woodland community in $kg\,m^{-2}$

Remember

★ feeding transfers food energy from one feeding level to the next

METABOLISM
Page 17.

★ energy is lost from each feeding level through life's activities, mostly in the form of heat released by the metabolism of cells.

As a result, the amount of food energy in a feeding level is less than the one below it.

As a result, the amount of living material (biomass) in a feeding level is less than the one below it. The diagram below gives you the idea.

Now you know why the number of links in a food chain is usually four or less (see page 26). When food energy dwindles to zero, feeding levels and links in food chains can no longer exist.

FACTS

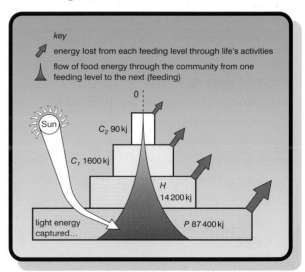

Pyramid of energy for a stream in $kJ\,m^{-2}\,yr^{-1}$

1.5 Population size

preview

preview

At the end of this section you will:

- **know how populations increase in size**
- **understand that the impact of human activity on the environment is related to population size**
- **be able to identify specific effects of human activity on the environment.**

The size of a population

★ A **population** is a group of individuals of the same species living in a particular place at the same time.

★ **Births** and **immigration** increase the size of a population.

★ **Deaths** and **emigration** decrease the size of a population.

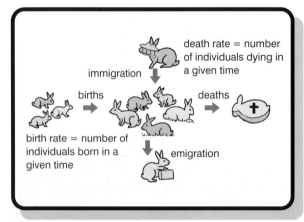

Factors affecting the size of a population

Population growth

The following graph shows that populations grow in a particular way.

Limiting factors stop populations from growing indefinitely. They include:

shortages of
- food
- water
- oxygen
- light
- shelter

build-up of
- poisonous wastes
- predators
- disease
- social factors.

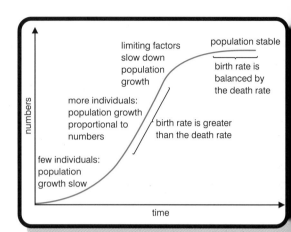

Population growth curve assuming immigration and emigration are in balance

Predator and prey populations

Predation affects the prey population. The number of prey affects the predator population: if prey is scarce, then some of the predators will starve. The graphs below show the relationships between the numbers of predators and prey.

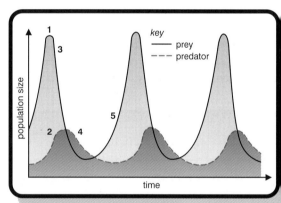

Predator–prey relationships

1 Prey breed and increase in numbers if conditions are favourable (e.g. food is abundant).

2 Predators breed and increase in numbers in response to the abundance of prey.

3 Predation pressure increases and the number of prey declines.

4 Predator numbers decline in response to the shortage of food.

Predation pressure decreases and so prey numbers increase ... and so on.

●tice that

● fluctuations in predator numbers are *less* than fluctuations in prey numbers

● fluctuations in predator numbers *lag* behind fluctuations in prey numbers.

ιy is this? There are fewer predators than ey, and predators tend to reproduce more ●wly than prey.

ne human population

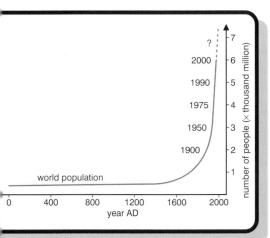

world population

Id population growth over the past 2000 years showing predicted re increase based on present trends

ιe graph shows that the human population has ●wn dramatically since the beginning of the ιeteenth century. Although the populations of ιrope, North America and Japan (developed ●untries) are levelling off, the populations of ιtin America, Asia and Africa (developing ●untries) are still growing rapidly as a result of

● improvements in food production
● more drugs for the treatment of disease
● improved medical care
● improved public health.

ιe rate of population growth is affected by the ●mber of young people in the population, ●rticularly women of child-bearing age. The ●le summarises the problem.

A large proportion of the world's population is young so the problems listed in the table are global. The problems are particularly acute in developing countries.

problem	result
present birth rate high	adding to the rate of population increase
future birth rate high	as children in the population 'bulge' grow older and have their own children, adding further to the rate of population increase
social services inadequate	large numbers of children put strain on the educational system, medical services and housing

Problems of a young population

In Britain (and other developed countries) the problems are more to do with a population that has an increasing proportion of old people. The diseases of old age (cancer, arthritis, dementia) take up an increasing proportion of the resources available for medical care.

1.6 Control and management

preview

At the end of this section you will:
● **know that the human impact on the environment is agricultural, industrial and domestic**

● **understand that pollution affects air, fresh water, sea water and land**

● **know about different ways of controlling pollution**

● **understand the advantages and disadvantages of using different energy sources.**

Today about 6 billion (six thousand million) people live in the world. Their impact on the environment is **agricultural, industrial** and **domestic**. Everyone needs food and a home. Many people also want a wide range of manufactured goods. In developed countries large quantities of food are produced and plenty of goods are manufactured. However, pollution is the result of plenty.

★ **Air** is polluted by gases, dust and smoke from cars, factories and houses.

★ **Fresh water** (rivers and lochs) and **seawater** are polluted by:
  • waste from factories
  • 'run off' of agrochemicals (pesticides and fertilisers) from farmland
  • discharge from sewage treatment works
  • 'cooling water' discharged from power stations.

★ **Land** is polluted by waste in the form of:
  • household (domestic) rubbish and waste from factories
  • pulverised ash from coke and coal-fired power stations
  • slag from quarrying and mining operations.

Controlling pollution

What can be done about reducing pollution?

Clean up the gases and liquids discharged into the environment

★ *Power stations* – 'scrubbers' in the chimneys of power stations remove sulphur from the waste gases produced by burning coal, before the gases escape into the atmosphere.

★ *Cars* – catalytic converters fitted to the exhausts of cars remove harmful gases (e.g. carbon monoxide) produced by engines burning fuel.

★ *Drinking water* – stripping nitrates and phosphates from drinking water reduces the risks to human health. The nitrates and phosphates are components of fertiliser which 'runs off' in solution from farmland into rivers and lochs.

Recycle materials

Recycling materials saves raw materials obtained from the environment and the energy needed to manufacture goods.

★ *Costs* of disposing of waste material in landfill sites are reduced.

★ *Space* which would be used for landfill sites is saved.

Reduce the use of harmful materials

★ *Chlorofluorohydrocarbons* (CFCs) are compound used as propellants in aerosols. They react with ozone in the upper atmosphere. Oxygen is produced and the ozone layer which protects life on Earth from the harmful effects of ultra-violet radiation is gradually destroyed. 'Ozone-friendly' compounds are replacing CFCs in a range of products.

Fossil fuels and nuclear power

Fossil fuels – coal, oil (including petrol) and gas – are a source of energy. Their combustion (burning) releases gases that pollute the environment. The table summarises the problem

pollutant	source	harmful effect
sulphur dioxide	power stations and factories	causes acid rain which damages trees and other plants. Acid rain also kills wildlife in lakes and ponds
carbon monoxide	engines and factories	combines with haemoglobin (see page 86) better than oxygen. A shortage of oxygen causes headaches and dizziness in affected people. A concentration of 0.1% of carbon monoxide in the air kills
oxides of nitrogen	engines, power stations, factories, mining	cause acid rain (see above). Increase the concentration of ozone in the atmosphere, damaging crops
Lead compounds	engines	poison the nervous system, causing damage to the brain. Replacing the lead in petrol with alternative substances and increasing use of *unleaded* petrol is reducing the amount of lead released into the environment

The main pollutants in air

clear reactors are also a source of energy. e safe disposal of the radioactive waste oduced by nuclear reactions is a major vironmental problem because it:

persists in the environment for thousands of years

accumulates in food chains (see page 26).

As a result human health is affected:
- the risk of the development of different types of cancer (e.g. leukaemia) increases
- women give birth to deformed babies.

.7 Water pollution

preview

preview

t the end of this section you will:
know that water is polluted from different sources

understand the effect of organic waste on the abiotic and biotic parts of a river ecosystem

be able to identify different indicator species.

What is the problem?

Living standards depend on industry and industry needs water to dispose of wastes. FARMING Pages 36–38.
Agrochemicals (pesticides and fertilisers) 'run off' from the land into rivers, lochs and the sea.
People produce sewage which enters (treated and untreated) rivers and the sea.

As a result, water is polluted making it:
- unsafe to drink (unless treated)
- unsuitable for organisms that depend on clean water to survive.

Organic waste consists of compounds containing carbon and originates from animals (including humans) and plants. It is a source of food (nutrients) for microorganisms. Sewage and fertilisers are examples. As the water becomes richer and richer in nutrients the microorganisms multiply and rapidly increase in numbers.

 FERTILISERS Page 37.

★ The activities of the microorganisms (including **aerobic** bacteria) decompose the organic waste

As a result the concentration of dissolved oxygen is reduced and poisonous substances (e.g. ammonia) are released into solution.

As a result organisms die through lack of oxygen and/or poisoning.

★ Anaerobic bacteria continue the decomposition of the organic waste

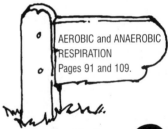 AEROBIC and ANAEROBIC RESPIRATION Pages 91 and 109.

As a result the water becomes black and unable to support wildlife

The sequence runs:

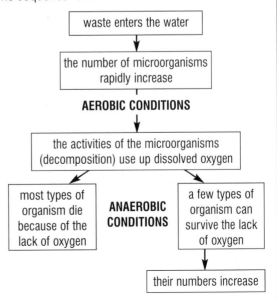

waste enters the water

↓

the number of microorganisms rapidly increase

AEROBIC CONDITIONS

↓

the activities of the microorganisms (decomposition) use up dissolved oxygen

| most types of organism die because of the lack of oxygen | **ANAEROBIC CONDITIONS** | a few types of organism can survive the lack of oxygen |

their numbers increase

The diagram shows you what happens to the animals of a river that is polluted with sewage and fertiliser.

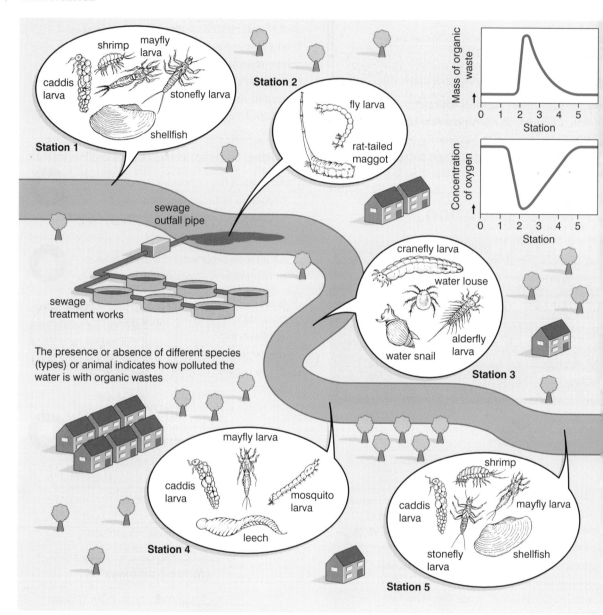

Notice that

★ the more waste there is in the water, the less oxygen there is in solution (abiotic effect).

ABIOTIC and BIOTIC Pages 19–20.

As a result there are fewer types of animal able to survive in the polluted parts of the river (biotic effect).

8 Managing the environment

preview

the end of this section you will:

understand how people exploit different environments

know about different ways of managing environments

understand how management can improve environments.

ferent pressures increase the use of land:

economic development

increasing human populations

the increasing need for food for the increasing population.

se study: tropical rain forest

e diagram shows you some of the ways people loit tropical rain forest.

Improvements

Managing tropical rain forests for human benefit and conserving the environment is possible.

★ Stop clearing rain forest

★ Use resources in a **sustainable** way. For example:
  • do not fell valuable hardwoods like teak and mahogany faster than they can be replaced by the reproduction and growth of new trees
  • subsidise poor countries so they will not need to make money from forest clearance
  • do not use rain forest products faster than they can naturally be replaced
    – rubber latex
    – nuts and fruits
    – plants as sources of new drugs and medicines.

Exploiting tropical rainforest

Rainforests girdle the equator covering 14.5 million km^2 of land. The vegetation recycles carbon dioxide and oxygen through photosynthesis.

Moisture absorbed by the forest evaporates back into the atmosphere, to fall as rain thousands of miles away. Rainforest is being cleared at a rate of 100 000 km^2 each year for:

cheap beef is exported to be made into hamburgers

Beef: about 20 000 km^2 of Brazilian forest are cleared each year for cattle ranches.

Opencast mining for metals causes much damage to the rainforest.

After clearing, nutrients disappear and the soil is soon exhausted. Semi-desert develops: the ranchers move on to clear a new area.

Logging: only 4% of trees are felled for timber, but another 40% are damaged or destroyed in the process.

Case study: fishing

The diagram shows you the human threat to stocks of commercially important fish.

Improvements

Managing fish stocks highlights the conflict between *short-term* human needs (food, economical use of expensive equipment) and the health of a particular environment. Using living things for food and as a source of different products only as fast as they can be naturally replaced helps the environment to provide resources in the *long term*.

★ Reducing the intensity of fishing helps balance the numbers of fish being born with the number of fish caught.

★ Increasing the mesh size of nets allows young fi[sh] to escape capture.

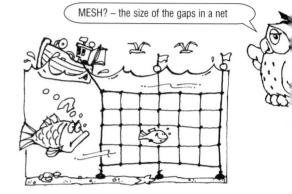

MESH? – the size of the gaps in a net

As a result more fish mature and reproduce to replace the fish caught.

★ Encourage the catching of little-used species

Fishing in the North Sea

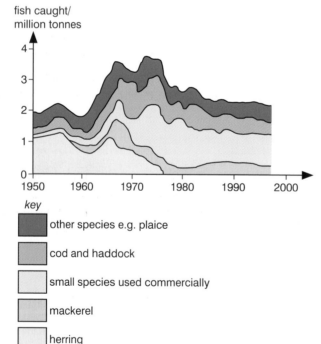

fish caught/ million tonnes

key

- other species e.g. plaice
- cod and haddock
- small species used commercially
- mackerel
- herring

Catches of fish are reduced because o[f]

★ **overfishing:** increased efficiency of fishing methods catches more fish than are replaced by reproduction

★ **pollution:**
- nutrients (e.g. nitrogen and phosphorus) from sewage works and surplus artificial fertilisers ente[r] rivers which discharge into the North Sea.
- pesticides used to protect crops enter rivers which discharge into t[he] North Sea
- metals (e.g. mercury, cadmium, copper) from different industrial processes

Case study: farming

Soil erosion

Erosion removes the top few centimetres of soil needed to grow crops. The soil may be washed away by rain or blown away by wind. Different human activities increase the risk of soil erosion.

★ **Clearing** land of trees and hedges

As a result the soil is no longer held in place by the roots of plants.

As a result there are no leaves to add humus which improves the structure and water-holding properties of soil.

Poor **farming** methods – for example:
- *overgrazing* of grassland by livestock (e.g. cattle and sheep)
- *cultivating* hillsides without taking account of the slope of the land.

Improvements

Growing crops and raising livestock is possible without damaging the soil.

Erosion by wind is prevented through

Reforestation, which replaces trees and hedgerows

As a result roots prevent soil from blowing away.

As a result humus is added to the soil.

As a result the soil remains fertile, maintaining crop yields.

Reducing the numbers of livestock grazing grassland

As a result the yield of grass needed to support livestock is maintained.

Erosion by running water is prevented through

Contour ploughing: furrows lie across the natural slope of the land.

Terracing: broad shallow ditches lie across the natural slope of the land.

As a result water is held back from flowing downhill.

As a result soil is *not* washed away.

Agrochemicals

Pests are plants and animals that destroy crops and livestock or prevent land from being used for farming. **Pesticides** are chemicals that kill pests.

Insecticides kill insects.

Herbicides kill weeds (plants that compete with crops for space, light and nutrients).

Fungicides kill fungi.

As a result food production increases.

Pesticides are very poisonous. Farmers apply pesticides to the land.

★ Pesticides can seep into ground water, contaminating drinking water and harming wildlife.

★ Pesticide spray can be carried in the wind and harm wildlife elsewhere.

Fertilisers supply nutrients which improve the growth of crops. Compounds of nitrogen (N), phosphorus (P) and potassium (K) – NPK fertilisers – are used in huge quantities.

POLLUTION
Pages 31–34.

★ Surplus NPK fertiliser runs off the land into rivers, lochs and the sea.

As a result water is polluted.

Improvements

★ Alternative methods of pest control (biological control) reduce dependence on pesticides.

★ Controlling application of pesticides helps to reduce the damage to wildlife and the environment.

★ Careful use of fertilisers reduces run off from the land.

Fact file

Before the introduction of pesticides, the management of food production depended on crop rotation: a different crop was grown in a particular field each year.

★ Each crop absorbed different nutrients from the soil.

As a result the crops' demands for nutrients are spread over a number of years.

★ Leguminous plants (e.g. clover) are part of the cycle of crop rotation.

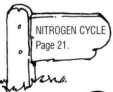

NITROGEN CYCLE
Page 21.

As a result soil nitrates are replaced by the nitrogen-fixing bacteria, which live in swellings (nodules) on the roots of leguminous plants.

Crop rotation prevents the build up of:

*   weeds
*   disease-causing organisms which affect crops

and maintans the fertility of the soil.

Today farmers use crop rotation and agrochemicals to maintain productivity.

Case study: forestry

Trees are planted as a resource in the future for making paper, construction and furniture. To be a useful product trees need time to grow to maturity.

★ **Conifers** (pine trees) take 70–80 years

★ **Hardwood** trees (e.g. oak, beech) take more than 100 years

As a result **forestry** (growing trees for commercial use) provides resources for human benefit in the long term.

Farmers and **foresters** (people who manage forests) have similar problems of environmental management. However, foresters also have other factors to consider.

★ Trees are often grown on poor quality soil which is exposed to severe weather conditions.

As a result young trees (**seedlings**) are grown in protected **nurseries** and then transplanted to the place where they grow to maturity.

★ Before planting, the ground may need deep ploughing and drainage.

★ Young trees need protection from animals that eat bark and young shoots (e.g. deer).

★ **Thinning** removes unhealthy trees and slow-growing trees which would otherwise compete for resources with the stronger specimens.

COMPETITION Page 26.

★ Precautions against fire protect trees from destruction during prolonged dry weather.

Finally: planting forests in places where trees would not normally grow may destroy wildlife. For example, planting forests in the Flow country of Caithness (an area of natural swampland and peat bog) may cause widespread damage. Species unique to the environment are driven to extinction and the habitat is unsuitable for the long-term survival of trees.

Words to remember

You have read some important words in this chapter. Here's a list to remind you what the words in green mean.

Adapted	suited for a particular job or t particular way of life in a particular environment
Biological control	the use of a species (usually a predator or parasite) to contro the numbers of another (pest) species
Community	all the organisms living in a particular environment
Compound	the substance formed when tw or more elements join togethe chemically
Decomposition	the breakdown physically and chemically of dead organisms living things. The process recycles nutrients (e.g. nitroge and carbon) through the environment
Emigration	movement of individuals out o population
Habitat	the place in the environment where a group of organisms liv
Humus	the dark fibrous material form by the decomposition of dead organisms
Immigration	movement of individuals into population

guminous   the family of plants (e.g. peas, beans, clover) which have root nodules containing nitrogen-fixing bacteria

trients   substances essential for growth and other living processes

ganism   any living thing

otosynthesis   the chemical reactions that use the light energy trapped by chlorophyll to convert carbon dioxide and water into sugars and oxygen

Pollution   occurs when substances produced by human activities are released into the environment, causing harm

Predation   the action of predators in catching food (prey)

Species   a group of individuals able to mate to reproduce offspring, which themselves are able to mate and reproduce

round-up

How much have you improved? Work out your improvement index on pages 134–135.

1 Look at pages 20–21. List the different components of an ecosystem. [6]

2 a) Explain the meaning of the word 'abiotic'. [1]
   b) Why is the amount of light an important abiotic influence on life inside a wood? [4]

3 Look at the pond food chain on page 26.
   a) How many links are there in the food chain? [1]
   b) Name the producers. [1]
   c) Briefly explain why they are called producers. [1]
   d) Name the herbivores. [1]
   e) Briefly explain why they are called herbivores. [1]
   f) Name the carnivores. [2]
   g) Briefly explain why they are called carnivores. [1]

4 The diagram shows a pyramid of biomass for a rocky seashore.

   a) Name the producers. [1]
   b) Name the secondary consumers. [1]

c) Severe weather conditions virtually wipe out the periwinkles. What will be the effect on the biomass of dog whelks and saw wrack? [2]

5.8 gm⁻²   dog whelks

71 gm⁻²   periwinkles

3987 gm⁻²   saw wrack

5 Look at the diagram on page 27 which shows the flow of energy through a food chain.
   a) How much of the energy which falls on $1m^2$ of plant in one year will pass on to the rabbit? Show your working. [3]
   b) Use the information on the diagram to explain why there are rarely more than four or five links in a food chain. [5]

Well done if you've improved. Don't worry if you haven't. Take a break and try again.

The world of plants

How much do you already know? Work out your score on page 135.

Test yourself

1 a) Name the inorganic substances that are the raw materials for photosynthesis. [2]
   b) Name the gas given off during photosynthesis. [1]

2 Complete the following paragraph using the words below. Each word may be used once, more than once or not at all.

transpiration evaporation increase translocation osmotic stomata xylem osmosis

Root hairs _____ the surface area available for the _____ uptake of water. _____ also transports water across the root into the _____ . Water moves through the _____ tissue in unbroken columns connecting the root with the leaves of the plant. Water is lost from the leaves by _____ through the _____ . [7]

3 Match the structure in column A with its correct description in column B. [5]

A	B
Nectary	Develops from the ovary after fertilisation
Stigma	A fertilised ovule
Fruit	Contains the egg nucleus
Ovule	Produces a sugar solution
Seed	Structure to which pollen grains attach

4 a) Briefly explain two different ways in which pollination occurs in plants. [4]
   b) What is the difference between pollination and fertilisation? [4]

5 Name the vegetative parts of a flowering plant. [3]

6 What are the advantages to growers of reproducing crops asexually? [3]

7 The list summarises the main stages in the production of whisky. Arrange the stages in the correct sequence.

drying production steeping grinding germination [5]

• •

2.1 Introducing plants

preview

At the end of this section you will:
● **be able to tell the difference between flowering plants and non-flowering plants**
● **know different uses of plants**
● **understand the possible consequences of a reduction in the variety of species of plants.**

There is a huge variety of plants. Some have flowers (**flowering plants**), others do not (**non-flowering** plants).

Notice that

★ flowering plants and conifers reproduce by means of **seeds**

SEEDS
Pages 47–48.

★ mosses and ferns reproduce by means of **spores**

★ ferns, conifers and flowering plants have **roots**. Moss plants do not have roots.

Fact file

Fungi and **algae** are sometimes described as plants. However, they are *not* plants.

★ The cells of fungi are quite unlike the cells of plants. They
   • form thread-like structures called **hyphae**
   • do *not* contain chlorophyll – an important featu of most plant cells.
★ Algae are either single-celled organisms or consi of many *similar* types of cell. Plants are made of many cells of *different* types.

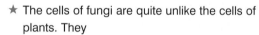

REPRODUCE BY MEANS OF SPORES

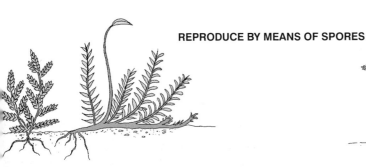

Mosses
- Mosses quickly lose water in dry air.
- As a result, mosses live in damp places.
- Roots are absent.
- As a result, water is soaked up by leaves.
- Stalks grow from moss plants, each carrying a capsule filled with spores.
- Each spore is able to grow into a new moss plant.

Ferns
- A waxy layer waterproofs the plant's surfaces, reducing water loss in a dry atmosphere.
- Roots draw water from the soil.
- Capsules containing spores grow beneath the leaves.
- Each spore is able to grow into a new fern plant.

REPRODUCE BY MEANS OF SEEDS

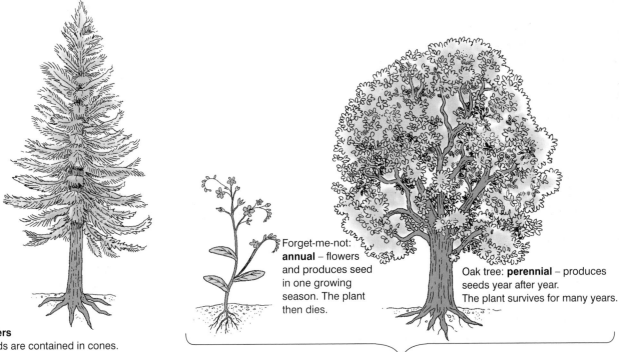

Forget-me-not: **annual** – flowers and produces seed in one growing season. The plant then dies.

Oak tree: **perennial** – produces seeds year after year. The plant survives for many years.

Conifers
- Seeds are contained in cones.
- Covered with leaves all year round **(evergreens)**.
- Roots draw water from the soil.
- Waxy layer waterproofs plant surfaces.

Flowering plants
- Seeds are contained in fruits.
- Leaves of trees/shrubs fall once a year **(deciduous)**.
- Roots draw water from the soil.
- Waxy layer waterproofs plant surfaces.

2

Using plants

The Mind Map on pages 44–45 shows some of the many different ways in which we use plants.

Notice that some uses of plants are *specialised* because the plants provide raw materials which are **processed** to make useful products.

★ **Malting barley** is used to make whisky and beer. During malting, starch in the grains of barley is converted into sugar. Yeast then ferments the sugar, producing ethanol (alcohol). The process runs:

FERMENTATION
Page 120.

★ **Rapeseed** contains oil which is used to make margarine, cooking oil and fats. The seeds are crushed to obtain the oil. The remains of crushed seeds are used in animal feeds.

★ **Raspberry** fruit separates from its core when picked, giving a 'hollow' berry. Most of the crop is preserved as jellies and sauces, canned or frozen. Enzymes can be used to extract the maximum volume of juice from raspberries.

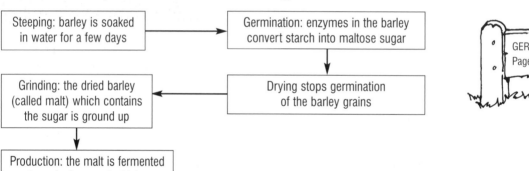

Steeping: barley is soaked in water for a few days	→ Germination: enzymes in the barley convert starch into maltose sugar
Grinding: the dried barley (called malt) which contains the sugar is ground up	← Drying stops germination of the barley grains
Production: the malt is fermented to make beer and whisky	

GERMINATION
Pages 48–49.

★ **Timber** is cut from trees and processed into wood products. The diagram gives you an idea of the variety of uses that are made of wood.

Slabs are the rounded sides of the log. They are broken up in a chipper and the pieces used to make chipboard and other wood products.

'Clear' lumber comes from the outer part of the log. It has the fewest knots (which weaken the timber) and is made into boards and planks

Hardwood from the centre of the log has more knots. It is cut into thick beams (large enough so that the knots do not weaken them) and used for house building and other construction work

Plywood is cut by rotating the log against a long blade. The wood peels off into thin layers.

Bark is stripped off and used for fuel and as soil conditioner

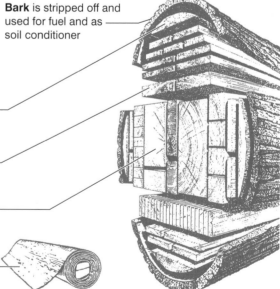

What is a knot?

Diagram adapted from Jarick, J. *et al*, *Plant Science – an introduction to world crops*, 3rd edition (W.H. Freeman & Co, New York).

Answer
Knots form where branches are removed during the life of a tree

xtinction

e huge variety of plants growing world-wide
kes possible the many different uses shown
the Mind Map on pages 44–45. However, the
mand for plant products and the clearance of
ests means that some plant species are on
e verge of extinction. The exploitation of
pical rainforests shown on page 35 illustrates
e problem.

tinction of plant species means that we are
prived of possible sources of new medicines
d food products. Also, varieties which help
prove the genetic quality of food crops may
lost.

member too that plants are at the
ginning of food chains and food webs.
eir loss threatens the consumers that
pend on them.

FOOD CHAINS
FOOD WEBS
Pages 26–27.

2.2 Growing plants

preview

At the end of this section you will:
- **know the different parts of flowers and understand their functions**
- **understand that pollination is the transfer of pollen from the male sex organs to the female sex organs**
- **understand that fertilisation is the fusion of a male sex nucleus with the female egg nucleus**
- **know that seed is formed from the fertilised egg.**

Flowers are shoots which are specialised for
sexual reproduction. Although flowers come in
different shapes and sizes, they are all made up
of similar parts: **sepals**, **petals**, **stamens** and
carpels. The parts of a flower and what each
part is for are shown in the diagram below.

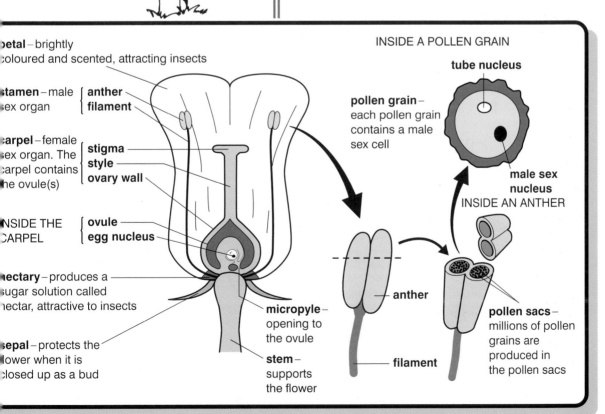

ts of an insect-pollinated flower

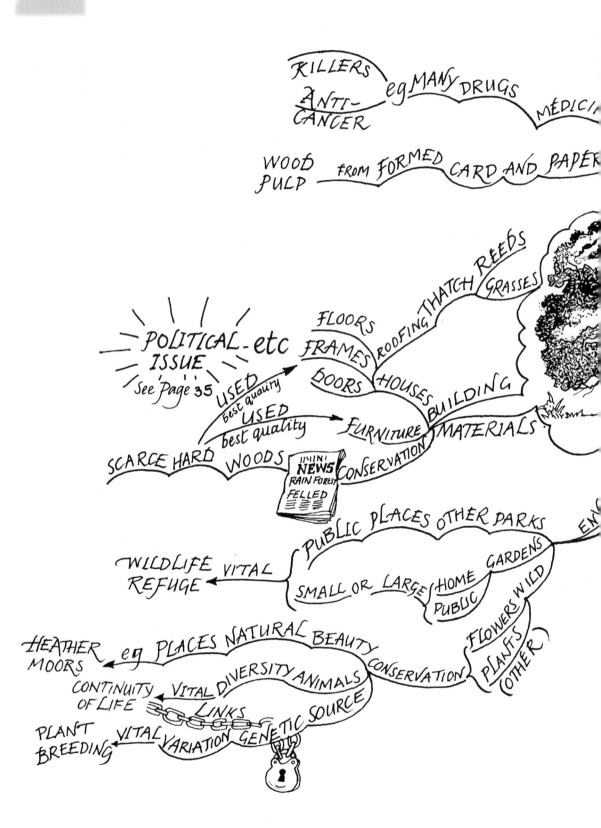

KILLERS
ANTI-CANCER

eg MANY DRUGS

MEDICI

WOOD PULP — FROM FORMED CARD AND PAPER

ROOFING THATCH REEDS GRASSES

FLOORS
FRAMES

POLITICAL - etc
ISSUE
See Page 35

DOORS HOUSES BUILDING

USED best quality
USED best quality

FURNITURE MATERIALS

SCARCE HARD WOODS

NEWS RAIN FOREST FELLED

CONSERVATION

PUBLIC PLACES OTHER PARKS EN

WILDLIFE VITAL REFUGE

SMALL OR LARGE HOME GARDENS
PUBLIC

FLOWERS WILD PLANTS (OTHER)

HEATHER MOORS
eg PLACES NATURAL BEAUTY CONSERVATION

CONTINUITY OF LIFE
VITAL DIVERSITY ANIMALS
LINKS GENETIC SOURCE

PLANT BREEDING
VITAL VARIATION

Some of the ways in which we use plants

FASHION!}

Dear Students,
The examples are only a
small selection of the foods
provided by plants

FOOD

ING
ANTS

FUEL

SALAD CROPS — SPRING ONIONS, LETTUCE, RADISH, BEETROOT

CEREALS — WHEAT, OATS, BARLEY, RICE

BREADS
BISCUITS and CAKE
FERMENTED DRINKS

VEGETABLES — ROOTS — CARROTS, TURNIPS, PARSNIPS, SWEDES

BRASSICAS — CABBAGE, CAULIFLOWER, BRUSSELS SPROUTS

FRUITS — RASPBERRIES, APPLES, TOMATOES, ORANGES, LEMONS

DRINKS — COCOA, COFFEE, TEA

WOOD AND CHARCOAL Produced by BURNING WOOD } in the absence of air (oxygen)

COAL FORMED from PLANTS FOSSILS

GASOHOL ETHANOL (ALCOHOL) PETROL MIXTURE produced by PLANT WASTE (eg sugar cane } FERMENTED

FERMENTATION
Page 120

45

Notice that the male sex cell is inside the **pollen grain** and that the female egg cell is inside the **ovule**.

Remember that **fertilisation** occurs when the nucleus of the male sex cell **fuses** with the nucleus of the female egg cell inside the ovule.

> How does tthe male sex nucleus reach the female egg nucleus?

★ **Pollination** is the process that brings pollen grains from the anthers to the **stigma** of a carpel. Insects are very important as pollen carriers.

> Keep studying the diagram of the flower on page 43.

★ Each pollen grain sprouts a **pollen tube**, which grows through the **style** of the carpel to the opening to the ovule. Find the opening on the diagram. The opening is called the **micropyle**.

★ The male sex nuclei pass through the micropyle into the ovule.

★ One of the male sex nuclei fuses with the female egg nucleus.

> **Now** you know how fertilisation takes place in flowering plants. The diagram below shows you the process.

And the other pollen tube nuclei?
At the end of its journey the tube nucleus dies, its job of controlling the growth of the pollen tube done. The male sex nucleus that does not fuse with the female egg nucleus fuses with other cell nuclei in the ovule.

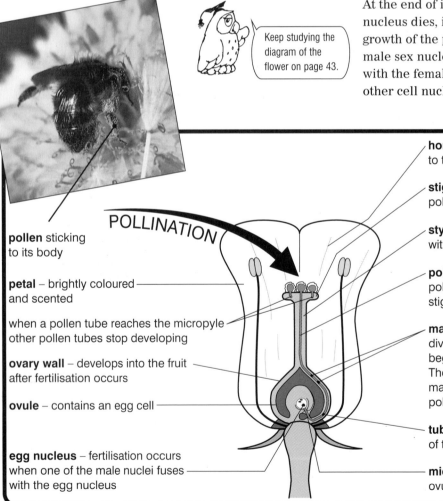

honey guides – point insects to the nectaries

stigma – pollination brings pollen grains to the stigma

style – connects the stigma with the ovary

pollen tube – grows from the pollen grain, down through the stigma to the micropyle

male nuclei – the male nucleus divides when the pollen grain begins to sprout a pollen tube. The two new nuclei are the male nuclei and pass down the pollen tube

tube nucleus – controls growth of the pollen tube

micropyle – opening to the ovule

POLLINATION

pollen sticking to its body

petal – brightly coloured and scented

when a pollen tube reaches the micropyle other pollen tubes stop developing

ovary wall – develops into the fruit after fertilisation occurs

ovule – contains an egg cell

egg nucleus – fertilisation occurs when one of the male nuclei fuses with the egg nucleus

Fertilisation follows pollination. The bee is visiting a flower. Pollen is carried by the bee in pollen sacs, one on each of its back legs: one pollen sac c be seen. Pollen also sticks to the bee's body

ore about pollination

ny do you notice flowers? Probably because
ey are brightly coloured and have a pleasant
ell. Insects notice flowers too (see the figure).
ey are attracted to

the **sweet smell** of many flowers

the **bright colours** of large petals

marks on the petals called **honey guides**

nectar, which is a sugar solution produced by the
nectaries at the base of the petals. Nectar is an
important food source for many types of insect. The
honey guides point insects in the direction of the
nectaries.

of these features of flowers attract insects
pecially bees).

en visiting flowers, insects pick up a load of
len, which is carried to the next flower they
it. Some of the pollen rubs off onto the
gma(s) of the carpel(s) and pollination takes
ce (see the diagram on page 46 to remind
rself of how this happens).

me flowers do not depend on insects for
lination. Instead they take advantage of the
nd to scatter pollen far and wide (see the table
ow).

Cross-pollination and self-pollination

When insects or wind transport pollen from the
anthers of the flower of one plant to the stigma(s)
of the flower of another similar plant, **cross-
pollination** has taken place. Pollen is transferred
between plants of the same type.

If pollen is transferred from the anthers to the
stigma(s) of the *same flower* or to the stigma(s) of
another flower on the *same
plant*, then **self-pollination**
has taken place. Cross-
pollination increases
genetic variation.

VARIATION
Pages 112–113

Seeds

A **seed** forms from the fertilised egg. It contains
the embryo plant (see page 48) and a **food store** of
starch. The embryo plant uses the stored food as a
source of energy during its germination and
growth into a new plant. The diagram shows
inside a seed of the broad bean.

The seed of a broad bean is **dicotyledonous**. It
contains **two** cotyledons (each a store of food).
Grass seed is **monocotyledonous**. It contains **one**
cotyledon.

art of flower	Insect-pollinated plants	Wind-pollinated plants
etals	• Brightly coloured • Usually scented • Most have nectaries	• If present, green or dull colour • No scent • No nectaries
nthers	• In a position where insects are likely to brush against them	• Hang loosely so that they shake easily in the wind
tigma	• In a position where insects are likely to brush against them • Sticky	• Feathery branches catch wind-blown pollen grains
ollen	• Small amounts produced • Large grains • Rough or sticky surface, which catches on the insects' bodies	• Large amounts produced • Small light grains • Smooth surface, which easily catches the wind

paring insect-pollinated flowers and wind-pollinated flowers

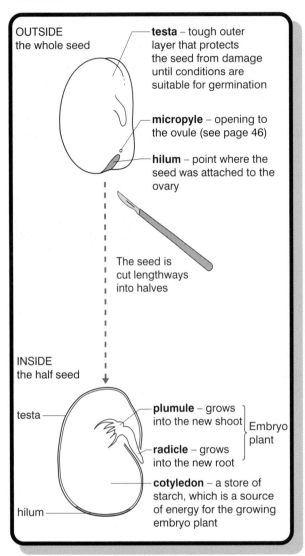

OUTSIDE
the whole seed

testa – tough outer layer that protects the seed from damage until conditions are suitable for germination

micropyle – opening to the ovule (see page 46)

hilum – point where the seed was attached to the ovary

The seed is cut lengthways into halves

INSIDE
the half seed

testa

plumule – grows into the new shoot ⎤ Embryo
radicle – grows into the new root ⎦ plant

cotyledon – a store of starch, which is a source of energy for the growing embryo plant

hilum

Outside and inside a broad bean seed

Seed germination

Germination describes the stages of growth from the embryo to the time when the seedling plant no longer depends on the food (starch) in the cotyledons.

★ **Water**: the seed absorbs water, softening the **testa** (seed coat) which splits. Growth of the embryo plant begins.

★ **Temperature**: warmth promotes germination. If the temperature is too low germination is prevented. The plant is safeguarded from frost and ice which could otherwise kill young seedlings.

The diagram shows the changes in percentage germination over a range of temperature. Temperature affects the enzymes controlling the chemical actions which bring about germination. If the temperature is too:

- *low*, the enzymes work slowly or not at all (deactivated)
- *high*, the enzymes may be destroyed (denatured).

The temperature at which germination is most successful (the optimum temperature) depends on the

- type of plant
- climate to which the plant is normally exposed

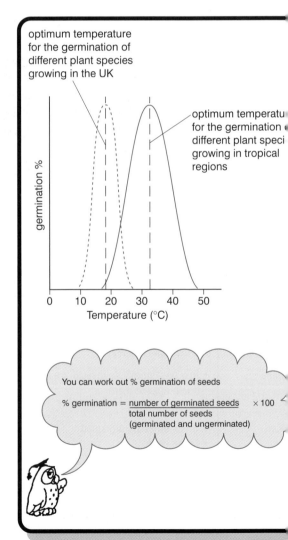

optimum temperature for the germination of different plant species growing in the UK

optimum temperatu for the germination different plant speci growing in tropical regions

germination %

Temperature (°C)

You can work out % germination of seeds

% germination = number of germinated seeds × 100
total number of seeds (germinated and ungerminated)

ct file

nts which germinate at 5–15 °C grow
l in the UK. Plants which germinate
30–40 °C grow best in tropical regions.

Oxygen: the seedling needs more oxygen than
when it was an embryo inside the seed. The
oxygen is used during
aerobic respiration, which
releases energy for cell
division and the processes
of growth.

AEROBIC
RESPIRATION
Pages 91–92.

uits

er fertilisation a **fruit** develops, usually from
e wall of the **ovary**. The wall may be 'fleshy' (e.g.
nato) or hard (e.g. hazel nut). The fruit contains
e seed. It helps scatter the seed far and wide.
e scattering of seed is called **dispersal**.

Why does seed dispersal help
plants survive?

spersal of fruits and seeds

uits and the seeds they contain are adapted
dispersal, usually by either animals or wind.

Spines (e.g. horse chestnut – the 'conker' is the
seed inside the prickly fruit) and **hooks** (e.g.
agrimony) attach the fruit to passing animals.

Animals are attracted to feed on **brightly
coloured** fruits – e.g. raspberries. A tough wall
(**seed coat**) protects the seeds from the digestive
juices in the animal's intestine. The seeds
eventually pass out in the animal's faeces.

Parachutes (e.g. dandelion) and **wings** (e.g.
sycamore) increase the surface area of fruits,
helping them travel in the wind.

Answer So that the plants which grow from the
seeds are not overcrowded.

2.3 Asexual reproduction preview

At the end of this section you will:
- be able to identify the organs of asexual
  reproduction in flowering plants
- know that cuttings, graftings and
  micropropagation are used by farmers and
  gardeners to produce many identical plants
- understand that asexual reproduction
  preserves desirable characteristics and so
  guarantees plant quality.

Asexual reproduction gives
rise to **genetically identical
individuals**. This process
passes on exact copies of the
parent's genetic material to
the daughter cells.

MITOSIS
Page 88.

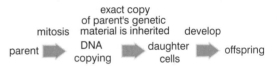

parent ➤ DNA copying ➤ daughter cells ➤ offspring

mitosis | exact copy of parent's genetic material is inherited | develop

Remember
★ Genetically identical individuals are called
  clones (see page 88).

Vegetative reproduction

Different parts of flowering plants can reproduce
asexually. They are called the **vegetative parts**
and are formed from the **root, leaf** or **stem**.

Asexual reproduction in flowering plants is
sometimes called **vegetative reproduction**. Since
the new plants come from a single parent, they
are genetically the same and are therefore clones.

The vegetative parts of plants store food. The
stored food is used for the development of the
new plant(s). We sometimes eat these food storage
organs as vegetables, for example, potatoes,
carrots.

The concept map for **asexual reproduction in
plants** is shown overleaf. It is your revision guide,
so study it carefully.

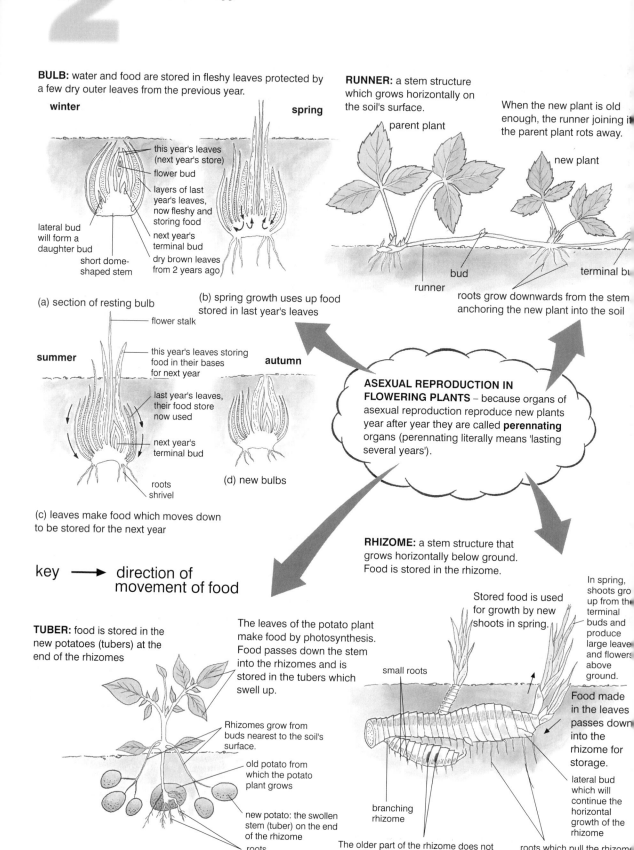

BULB: water and food are stored in fleshy leaves protected by a few dry outer leaves from the previous year.

winter

spring

this year's leaves (next year's store)

flower bud

layers of last year's leaves, now fleshy and storing food

lateral bud will form a daughter bud

next year's terminal bud

short dome-shaped stem

dry brown leaves from 2 years ago

(a) section of resting bulb

(b) spring growth uses up food stored in last year's leaves

summer

flower stalk

this year's leaves storing food in their bases for next year

autumn

last year's leaves, their food store now used

next year's terminal bud

roots shrivel

(d) new bulbs

(c) leaves make food which moves down to be stored for the next year

key ⟶ direction of movement of food

RUNNER: a stem structure which grows horizontally on the soil's surface.

When the new plant is old enough, the runner joining it the parent plant rots away.

parent plant

new plant

bud

terminal bu

runner

roots grow downwards from the stem anchoring the new plant into the soil

ASEXUAL REPRODUCTION IN FLOWERING PLANTS – because organs of asexual reproduction reproduce new plants year after year they are called **perennating** organs (perennating literally means 'lasting several years').

RHIZOME: a stem structure that grows horizontally below ground. Food is stored in the rhizome.

Stored food is used for growth by new shoots in spring.

In spring, shoots gro up from the terminal buds and produce large leave and flowers above ground.

TUBER: food is stored in the new potatoes (tubers) at the end of the rhizomes

The leaves of the potato plant make food by photosynthesis. Food passes down the stem into the rhizomes and is stored in the tubers which swell up.

small roots

Food made in the leaves passes down into the rhizome for storage.

Rhizomes grow from buds nearest to the soil's surface.

old potato from which the potato plant grows

new potato: the swollen stem (tuber) on the end of the rhizome

roots

branching rhizome

lateral bud which will continue the horizontal growth of the rhizome

roots which pull the rhizome into the soil (called contract

The older part of the rhizome does not die and shrivel for several years, so scars of the shoots from previous years can be seen along it.

tificial vegetative reproduction

·deners and farmers need to produce fresh
·ks of plants that have desirable
·racteristics such as disease resistance, colour
·ruit or shape of flower. The diagram below
·ws how they use vegetative propagation.

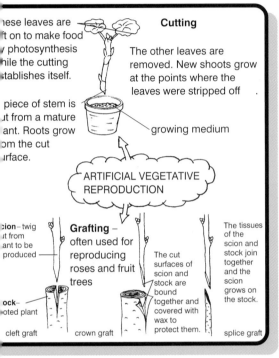

Cutting

·ese leaves are
·t on to make food
·y photosynthesis
·hile the cutting
·stablishes itself.

The other leaves are
removed. New shoots grow
at the points where the
leaves were stripped off .

·piece of stem is
·t from a mature
·ant. Roots grow
·om the cut
·urface.

growing medium

ARTIFICIAL VEGETATIVE
REPRODUCTION

·cion – twig
·ut from
·ant to be
·produced

Grafting –
often used for
reproducing
roses and fruit
trees

The cut
surfaces of
scion and
stock are
bound
together and
covered with
wax to
protect them.

The tissues
of the
scion and
stock join
together
and the
scion
grows on
the stock.

·ock–
·oted plant

cleft graft   crown graft   splice graft

·biting vegetative reproduction

·cropropagation is used to grow plants from
·all pieces, using a technique called **tissue
·ture**.

·Small fragments of plant tissue are grown in a
·iquid or gel that contains all the necessary
·ngredients.

·Conditions are sterile.

·As a result, the new plants are free
·of disease.

·The temperature is carefully controlled.

All the plants grown from pieces of one parent
plant will be genetically identical. They are
clones. The advantages are that the plants
* are healthy
* are the same
* retain the desirable characteristics of the
  parent plant.

Asexual or sexual?

During asexual reproduction, the cells of one
parent plant divide by **mitosis** to produce
daughter cells which form new individuals. The
offspring are genetically *identical* to one
another (**clones**) because DNA copies itself
during mitosis. The table below compares the
advantages and disadvantages of asexual
reproduction.

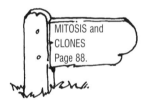

MITOSIS and
CLONES
Page 88.

COMPETITION
Page 26.

advantages	disadvantages
• successful characteristics are inherited unchanged by offspring from the parent plant	• there is little genetic variation between individuals, limiting adaptation to a changing environment
• offspring grow rapidly As a result offspring can rapidly colonise the environment	• overcrowding and competition for resources may occur, because offspring grow near to the parent plant.

In sexual reproduction, *two* parent plants each
produce sex cells (**gametes**). The male sex cell
fuses with the female sex cell (**fertilisation**).
The fertilised egg is genetically different from
either parent because of the contribution of
genetic material (**genes**) in
the gametes from each
parent. In other words,
the offspring **varies** from
its parents.

GENES
Page 115.

The table below compares the advantages and disadvantages of sexual reproduction.

advantages	disadvantages
• genetic variation allows individuals to adapt to changing environments	• different stages pollination → fertilisation → dispersal mean that the process is rather chancy
• dispersal occurs because seeds which are the results of sexual reproduction in plants are carried away from the parent plants.	• seeds may not land in environments suitable for growth and survival

Many different types of plant are able to reproduce asexually and sexually. For example, strawberries reproduce asexually from runners and sexually when they come into flower. They benefit, therefore, from both forms of reproduction.

2.4 Making food

and mineral nutrients

At the end of this section you will:
- **understand that leaves are adapted for photosynthesis**
- **know that limiting factors affect the rate of photosynthesis**
- **be able to identify the uses of sugar in plants.**

Photosynthesis

Photosynthesis is a chemical process that traps the energy of sunlight and uses it to convert carbon dioxide and water into sugars (food). A summary of the process is:

$$\text{carbon dioxide} + \text{water} \xrightarrow{\text{catalysed by chlorophyll}} \text{glucose} + \text{oxygen}$$
$$\text{(a type of sugar)}$$

There are lots of different chemical reactions that make up the process of photosynthesis. The reactions happen inside chloroplasts, in the leaves and other green parts of plant cells.

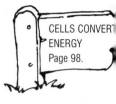
CELLS CONVERT ENERGY
Page 98.

Leaves

A leaf is a food-making factory. It is **adapted** for photosynthesis.

Carbon dioxide and water circulate within the leaf. Light is captured by the pigment chlorophyll which is packaged in the chloroplasts that pack the cells of the leaf. On pages 00–00 is the concept map for **photosynthesis**, and its checklist of points.

The diagram below shows you how the plant uses sugars.

The uses of sugar in plants

Investigating photosynthesis

Starch is a food substance stored in plant cells. Plants make sugar by photosynthesis, as we have seen. The sugar molecules are building blocks which combine to make starch. You can investigate photosynthesis by testing leaves for the presence of starch. The sequence runs:

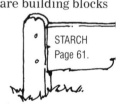
STARCH
Page 61.

★ **Destarching**: a potted plant (e.g. geranium) is placed in the dark (e.g. a cupboard) for several days. Any starch in the leaves is used up.

★ **Light**: some of the leaves of the plant are each covered with a pocket of black paper, shielding

hem from light. The other leaves are left uncovered
nd illuminated for several days.

reatment: leaves – covered and uncovered – are
detached from the plant and treated as follows:

killing and softening – the leaves are boiled in
water

bleaching – the leaves are warmed in ethanol
(**caution** – ethanol easily catches light (is
flammable)) removing chlorophyll. The leaves
are creamy white, making it easier to see the
results of the test for the presence of starch.

Softening: the leaves are brittle following treatment
vith ethanol. Placing them in water softens them
gain.

Testing: the leaves are covered with iodine
solution. Iodine solution is amber (orange/brown) in
colour. It changes to blue/black in the presence of
starch.

uncovered leaves are stained blue/black,
showing that starch is present in the leaves.

covered leaves are stained amber, showing that
starch is *not* present in leaves.

clusion: plants need light to make starch.
ne plants have variegated leaves. The diagram
ws that 'variegated' means that not all of each
f contains chlorophyll.

gated leaves which have been illuminated and treated as
cribed above are stained with iodine solution. Parts of the
where chlorophyll is present: stained blue/black. Parts of the
where chlorophyll is *not* present: stained amber
resentend as green in the diagram). *Conclusion:* starch is
uced only in areas of the leaf where there is chlorophyll.

Limiting factors

The rate at which plants make sugar by
photosynthesis is affected by supplies of **carbon
dioxide** and **water**, **temperature** and the
intensity of light. These factors are called
limiting factors because if any one of them falls
to a low level, photosynthesis slows down or
stops. The diagram below illustrates the point.
At low concentrations of carbon dioxide, the
carbon dioxide limits the rate of photosynthesis,
whatever the light level is. Carbon dioxide is
the limiting factor. At higher concentrations of
carbon dioxide, the rate of photosynthesis
increases if the light is bright enough. Light is
now the limiting factor.

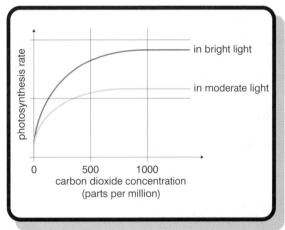

Effect of carbon dioxide concentration on the rate of photosynthesis

The higher the temperature, the faster the
chemical reactions of photosynthesis, within
limits. Extreme cold (below 0°C) deactivates
and extreme heat (above 45°C) denatures the
enzymes which control the chemical reactions
of photosynthesis.

Water is a raw material for photosynthesis.
However, water is also the solvent in which the
reactions of metabolism occur within cells.
Singling out the direct effect of the availability
of water on photosynthesis is therefore very
difficult.

In a greenhouse, conditions are controlled so
that limiting factors are eliminated.

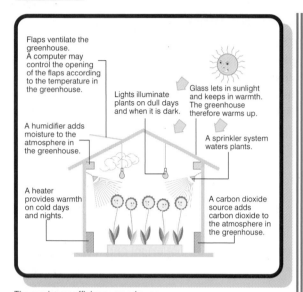

The maximum-efficiency greenhouse

Checklist for photosynthesis

1 ★ Leaves are arranged so that the lower ones are not overshadowed by those above. The arrangement is called the **leaf mosaic**.

As a result, more leaves are exposed to direct sunlight and take up more light.

★ The leaf blade is flat.

As a result, a large surface area is exposed for the absorption of light.

★ The leaf blade is thin.

As a result, light reaches the lower layers of cells in the leaf and gas reaches all the parts.

TRANSPIRATION
Page 58.

2 ★ Water moves up to leaves in the **transpiration stream**.

★ Carbon dioxide enters leaves 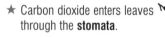 through the **stomata**.

3 ★ The cells of the upper leaf surface do not contain chloroplasts and are transparent.

★ **Palisade cells** beneath the upper epidermis are column shaped, tightly packed and filled with chloroplasts.

As a result, many chloroplasts are exposed to bright light, maximising the rate of photosynthesis.

★ **Spongy mesophyll cells** contain fewer chloroplasts and are more loosely packed.

As a result, there are air spaces between the spongy mesophyll cells.

★ The spaces allow carbon dioxide and water vapour to circulate freely within the leaf, bringing the raw materials for photosynthesis to the leaf cells.

STOMATA
Page 58.

★ Each **stoma** is flanked by guard cells which cont[...]
the size of the opening of the stoma.

As a result, the rate of diffusion of gases into and out of the leaf through the stomata is controlled.

★ The cells of the lower leaf surface lack chloroplast[...]
except the guard cells.

4 ★ Chloroplasts pack the inside of each palisade cell.

★ Chloroplasts stream in the cytoplasm (**cyclosis**) tc[...]
region of the palisade cell where light is brightest.

As a result, the rate of photosynthesis is maximised.

★ The chloroplasts are filled with the green pigment **chlorophyll**.

★ Chlorophyll absorbs light, especially wavelengths i[...]
the red and blue parts of the spectrum.

★ During photosynthesis, light energy is converted ir[...]
the energy of chemical bonds of glucose (sugar).

★ Chlorophyll makes light energy available for the synthesis (making) of sugar.

PHOTOSYNTHESIS IN ACTION
Water molecules and carbon dioxide molecules combine to form sugar.
The chemical reactions produce oxygen.

carbon dioxide + water ⟶ sugar + oxygen

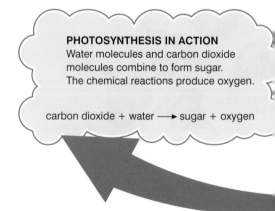

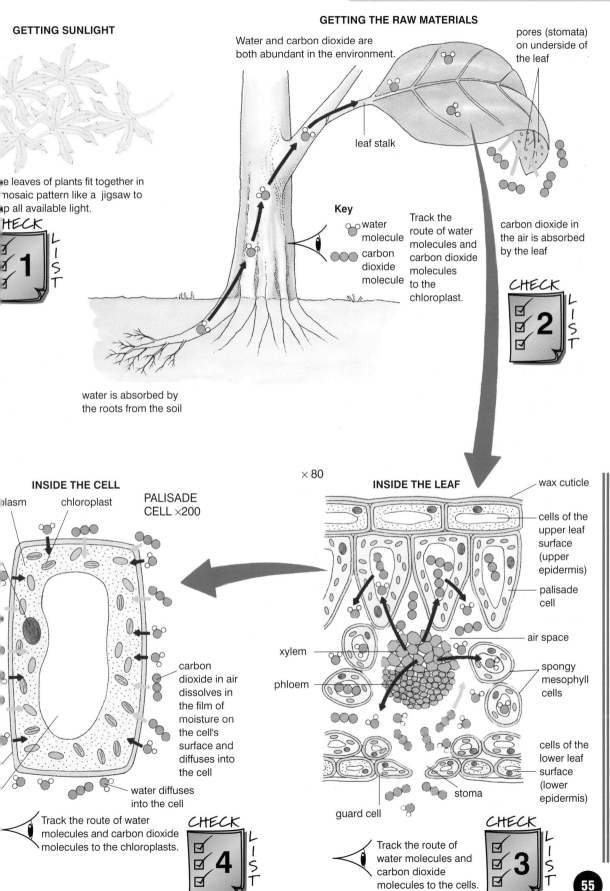

GETTING THE RAW MATERIALS

GETTING SUNLIGHT

Water and carbon dioxide are both abundant in the environment.

pores (stomata) on underside of the leaf

leaf stalk

...e leaves of plants fit together in ...osaic pattern like a jigsaw to ...p all available light.

Key

water molecule

carbon dioxide molecule

Track the route of water molecules and carbon dioxide molecules to the chloroplast.

carbon dioxide in the air is absorbed by the leaf

CHECK 1 LIST

CHECK ☑☑☑ 2 LIST

water is absorbed by the roots from the soil

× 80

INSIDE THE CELL

...plasm   chloroplast

PALISADE CELL ×200

INSIDE THE LEAF

wax cuticle

cells of the upper leaf surface (upper epidermis)

palisade cell

air space

xylem

phloem

spongy mesophyll cells

carbon dioxide in air dissolves in the film of moisture on the cell's surface and diffuses into the cell

water diffuses into the cell

cells of the lower leaf surface (lower epidermis)

guard cell

stoma

Track the route of water molecules and carbon dioxide molecules to the chloroplasts.

CHECK ☑☑☑ 4 LIST

Track the route of water molecules and carbon dioxide molecules to the cells.

CHECK ☑☑☑ 3 LIST

TRANSPORT OF FOOD

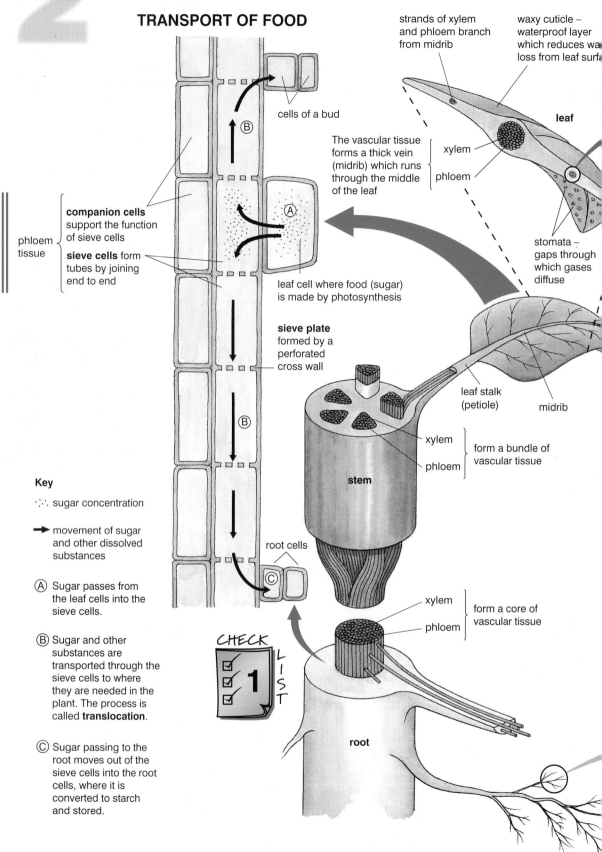

strands of xylem and phloem branch from midrib

waxy cuticle – waterproof layer which reduces wa loss from leaf surf

leaf

xylem

phloem

stomata – gaps through which gases diffuse

cells of a bud

The vascular tissue forms a thick vein (midrib) which runs through the middle of the leaf

companion cells support the function of sieve cells

phloem tissue

sieve cells form tubes by joining end to end

leaf cell where food (sugar) is made by photosynthesis

sieve plate formed by a perforated cross wall

leaf stalk (petiole)

midrib

xylem

phloem

form a bundle of vascular tissue

stem

root cells

Key

∴ sugar concentration

➤ movement of sugar and other dissolved substances

Ⓐ Sugar passes from the leaf cells into the sieve cells.

Ⓑ Sugar and other substances are transported through the sieve cells to where they are needed in the plant. The process is called **translocation**.

Ⓒ Sugar passing to the root moves out of the sieve cells into the root cells, where it is converted to starch and stored.

CHECK LIST 1

xylem

phloem

form a core of vascular tissue

root

TRANSPORT OF WATER

upper leaf surface

xylem

G

phloem

D

E

air space

F

lower leaf surface

guard cell

stoma

TRANSPIRATION

Xylem cells join end to end. Cross walls separating a cell from neighbouring cells are broken down forming a continuous tube, rather like a drinking straw.

CHECK
☑
☑ **2**
☑
LIST

→ movement of water

Root hairs absorb water from the soil – by **osmosis** – and mineral ions

Water passes through the root tissue into the xylem by **osmosis**.

Water travels through the xylem of the root and stem in unbroken columns – the **transpiration stream**.

Water moves through the xylem of the leaf stalk and veins of the leaf.

Water evaporates into the large air spaces within the leaf. The air spaces are saturated with water vapour. The concentration of water vapour in the atmosphere is lower than that in the air spaces. Water vapour therefore diffuses from the leaf through the stomata. The process is called **transpiration**.

Water lost by cells through evaporation is replaced with water drawn through the cells by osmosis. Cells next to the xylem draw water from the xylem by osmosis.

C

B

A

root hair cell

root cells

soil particles

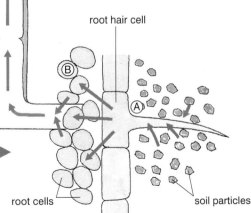

Checklist for transport in plants

1 ★ Phloem consists of living cells.

★ The concentration of sugar in the leaf is often lower than the concentration of sugar in the upper ends of the sieve tubes.

★ The sieve tubes lose sugar when it is stored as starch.

2 ★ Xylem consists of dead cells.

★ The walls of xylem tubes are waterproofed with a substance called **lignin**, which also gives extra strength, keeping the plant upright (**support**).

★ As water transpires, more is drawn from the xylem in the leaf.

As a result, water is 'sucked' upwards through the xylem of the stem.

As a result, more water is supplied to the bottom of the xylem by the roots.

As a result, there are unbroken moving columns of water from the roots to the leaves.

2

2.5 Transport in plants

preview

At the end of this section you will:
- **know that xylem tissue transports water and that phloem tissue transports food**
- **understand the processes of transpiration and translocation**
- **know that xylem tissue and phloem tissue form vascular bundles which reach all parts of the plant.**

Transport systems in plants

On pages 56–57 is the concept map for **transport in plants**, and its checklist. Study them carefully.

Factors affecting transpiration

Think of the ways plants lose and gain water.

★ Loss of water is through transpiration.

★ Gain is through the uptake of water by the roots.

If the loss of water is greater than the gain, then the stomata close.

As a result, transpiration is reduced.

If the loss of water is still more than the gain, then the cells of the plant lose turgor and the plant **wilts**.

The graphs show the effect of other factors on the rate of transpiration. Light stimulates the stomata to open wide. The rate of transpiration is therefore greater during the day than at night.

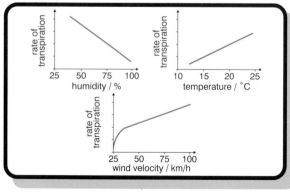

Factors affecting the rate of transpiration

Controlling the size of stomata

Two sausage-shaped **guard cells** surround the opening which forms the stoma. The guard cel contain chloroplasts. Think of the sequence:

★ During the day, photosynthesis increases the concentration of sugar in the guard cells. There a net flow of water by osmosis into the guard cells making them turgid. The guard cells bow out, opening the stoma.

OSMOSIS
Pages 56–57

★ At night, photosynthesis stops and the concentration of sugar in the guard cells falls. Th is a net outflow of water and the guard cells lose turgor. The guard cells bow in, closing the stoma

Words to remember

You have read some important words in this chapter. Here's a list to remind you what the words in green mean.

Clones	a group of genetically identic individuals
Embryo	the early stages of developme after fertilisation
Extinction	occurs when a species dies o
Ferments	describes the activities of cel and tissues producing differe substances as a result of the reactions of anaerobic (witho oxygen) respiration
Germination	the stages of growth from the embryo plant to the time whe the growing plant no longer depends on stored food
Photosynthesis	the chemical reactions that u the light energy trapped by chlorophyll to convert carbon dioxide and water into sugars and oxygen
Seedling	a young plant
Wilt	drooping of a plant through la of water

round-up

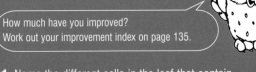

How much have you improved?
Work out your improvement index on page 135.

1 Name the different cells in the leaf that contain
  chloroplasts. [3]

2 Briefly explain why most chloroplasts are found
  in palisade cells lying just beneath the upper
  surface of the leaf. [4]

3 List the major limiting factors for photosynthesis. [4]

4 Complete the following paragraph using the words
  below. Each word may be used once, more than
  once or not at all.

**xylem photosynthesis transpiration
translocation**

Leaves produce sugar by _____.
The transport of sugar is called _____ .
The transport of water is called _____ . [3]

5 Describe the probable climate on a day when the
  transpiration rate of a plant is at a maximum. [4]

6 Briefly explain what happens if a plant loses
  more water through transpiration than it gains
  through absorption of water by its roots. [3]

7 Compare the characteristics of xylem tissue
  with those of phloem tissue. List the comparisons
  in two columns headed 'xylem' and 'phloem'
  respectively. [5]

Well done if you've improved. Don't worry
if you haven't. Take a break and try again.

Animal survival

How much do you already know?
Work out your score on page 135.

3.1 The need for food

preview

At the end of this section you will:
- **know that nutrients are the substance in food needed for a healthy life**
- **understand more about carbohydrates, lipid and proteins**
- **know why food is needed.**

Test yourself

1 The nutrients in food are listed below. Use these nutrients to answer the following questions.

carbohydrates   fats   proteins   vitamins   minerals

a) Which nutrients give food its energy content?   [3]
b) Which nutrient is a source of energy, but is most important in the body for growth and repair?   [1]
c) Which nutrient releases the most energy per gram?   [1]
d) Which nutrients are needed only in small amounts, but play an important role in the control of metabolism?   [2]

2 Match each term in column **A** with its correct description in column **B**.

A terms	B descriptions
ingestion	the removal of undigested food through the anus
digestion	digested food passes into the body
absorption	food is taken into the mouth
egestion	food is broken down [4]

3 The structures of the kidney tubule and its blood supply are listed below. Rewrite them in the order in which a molecule of urea passes from the renal artery to the outside of the body.

**tubule   urethra   bladder   glomerulus
Bowman's capsule   ureter   collecting duct**   [7]

4 How are the teeth of a dog adapted for grasping and cutting food?   [4]

• •

The **nutrients** in food are **carbohydrates**, **fats**, **proteins**, **vitamins** and **minerals**. **Water** and **fibre** are also components of food. Different foods contain nutrients, water and fibre in different proportions. Our **diet** is the food and drink we take in. Remember the sequence:

$$\left.\begin{array}{l}\text{nutrient} \\ +\ \text{water} \\ +\ \text{fibre}\end{array}\right\} \xrightarrow{\text{components of}} \text{food} \xrightarrow{\text{eaten}} \text{di}$$

All living things (including us) need food.

- **carbohydrates**: a major source of energy and structural materials
- **lipids**: stores of energy
- **proteins**: for building bodies

Elements for life

All matter is made of chemical elements. Of these elements, six make up more than 95% by mass of living matter. They are:

- carbon (C)
- hydrogen (H)
- nitrogen (N)
- oxygen (O)
- phosphorus (P)
- sulphur (S).

The symbols of the elements arranged in order of abundance in living matter make the mnemonic **CHNOPS**.

Hint & Tips

Carbon is the most common element in the substances that make up living things. Carbon atoms can combine to form long chains. Many of the carbon compounds in living things

…ve large molecules (**macromolecules**) formed by …all molecules combining.

…arbohydrates

…rbohydrates are compounds containing the …ements carbon, hydrogen and oxygen.

…nosaccharides are simple sugars. Sweet-tasting …ctose and **glucose** are examples. Both have the …rmula $C_6H_{12}O_6$. The six …rbon atoms form a ring. …gars (especially glucose) … an important source of …ergy in all living things.

fructose   glucose

formula of fructose and glucose in shorthand form

…saccharides are more complex sugars. They are …med when two monosaccharides combine. For …ample, two molecules of glucose combine to …m one molecule of **maltose**:

…2 glucose → maltose + water

$2C_6H_{12}O_6 \rightarrow C_{12}H_{22}O_{11}(aq) + H_2O$

The formula for maltose in shorthand form

…molecule of fructose and a molecule of glucose …mbine to form one molecule of **sucrose**:

…glucose + fructose → sucrose + water

…lysaccharides are carbohydrates whose …olecules contain hundreds of sugar rings. For …ample, **starch**, **cellulose** and **glycogen** are …lysaccharides. Their molecules consist of long …ains of glucose rings.

Starch is a food substance stored in plants. Their cells convert the starch into glucose, which is oxidised (respired) to release energy.

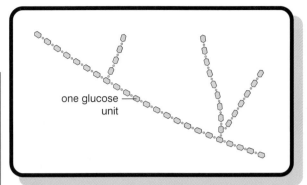

Part of a starch molecule

★ **Glycogen** is a food substance stored in animals. Liver cells convert glycogen into glucose, which is oxidised (respired) to release energy.

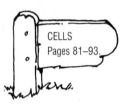

CELLS Pages 81–93.

★ **Cellulose** is an important component of the cell walls of plants.

Lipids

Lipids are compounds containing the elements carbon, hydrogen and oxygen. There are two types of lipids: **fats**, which are solid at room temperature; and **oils**, which are liquid at room temperature.

Fats and oils are compounds formed between two constituents: **fatty acids** and **glycerol**. A molecule of glycerol can combine with three fatty acid molecules to form a **triglyceride** molecule and three molecules of water. Fats and oils are mixtures of triglycerides.

glycerol + fatty acid $\longrightarrow$ triglyceride + water

$$\begin{array}{l} \text{OH} \\ \text{OH} + 3HA \\ \text{OH} \end{array} \longrightarrow \begin{array}{l} \text{A} \\ \text{A} \\ \text{A} \end{array} + 3H_2O$$

Making a triglyceride

Fats and oils are important as

* components of cell membranes
* sources of energy
* sources of the fat-soluble vitamins A, D and E
* insulation which helps keep the body warm
* protection for delicate organs.

Proteins

Proteins are compounds containing the elements carbon, hydrogen, oxygen, nitrogen and sometimes sulphur.

Amino acids are the building blocks which combine to make proteins. Two or more amino acids can combine to form a **peptide**, which can combine with more amino acids to form a **protein**.

Fact file

★ **Peptides** have molecules with up to 15 amino acids.

★ **Polypeptides** have molecules with 15–50 amino acids.

★ **Proteins** have still larger molecules.

There are 20 different amino acids that combine to form proteins. The protein made depends on the type, number and order in which amino acids join together.

Proteins are important because

- they are the materials from which new tissues are made during growth and repair
- **enzymes** are proteins which control the rates of chemical reactions in cells
- **hormones** are proteins which control the activities of organisms.

HORMONES
Page 75.

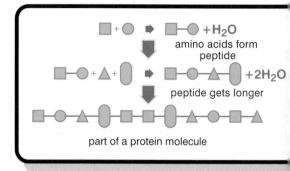

How amino acids combine to form peptides and proteins. Each shape represents a particular type of amino acid.

On page 6 is a Mind Map for **food and diet**. The fact file below gives you more information.

Fact file

★ A **balanced diet** is a mixture of foods which together provide sufficient nutrients for healthy living.

★ The 'basic four' food groups help us choose a balanced diet. The diagram gives you the idea.

★ Vitamin C helps cells to join together. It also controls the use of calcium by bones and teeth.

★ Vitamin D helps the body to absorb calcium.

★ Deficiency of iron is a common cause of anaemia.

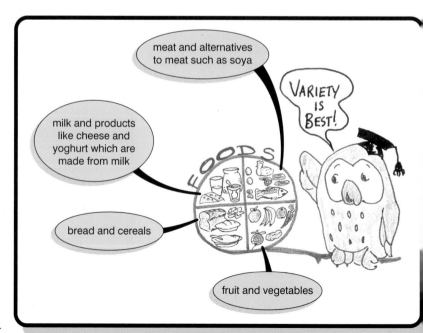

The 'basic four' food groups help us to choose a balanced diet. Eat at least one helping from each group daily. Choose different foods from each group for variety.

2 The digestive system

preview

t the end of this section you will:

know that the digestive system is a muscular tube through which food moves

understand that as food moves through the digestive system it is processed (digested) into substances which the cells of the body can absorb and use

be able to identify enzymes responsible for digesting food

know that digestive systems are adapted to type of diet.

sting your understanding
e terms:

t

estine

mentary canal

all refer to the digestive system.

gesting food

od is processed through the digestive system in e following sequence:

ingestion
food is taken
into the mouth
↓

digestion
large molecules of food which
the body cannot absorb are
broken down into smaller molecules
↓

absorption
the small molecules
of digested food
pass into the body
↓

egestion
undigested food is
removed from the body
through the anus

e digestive system is a muscular tube through ich food moves. It processes food.

★ **Mechanical processes** break up food and mix it with digestive juices.

★ **Chemical processes** digest food using different enzymes in the digestive juices. The body cannot absorb the large molecules of carbohydrate, protein and fat in food. They are broken down into smaller molecules which the body can absorb.

On pages 64–65 is the concept map for **the digestive system**. The **liver** and **pancreas** are connected to the digestive system. They play an important role in the digestion of food. The numbers on the concept map refer to the checklist below.

Checklist for the digestive system

(M) = mechanical processes of digestion

(C) = chemical processes of digestion

1 ★ **(M) Teeth** chew food, breaking it into small pieces.

As a result, the surface area of food exposed to the action of digestive enzymes is increased.

As a result, food is digested more quickly.

2 ★ **(C) Saliva**, produced by the salivary glands, contains the enzyme **amylase**.

As a result, the digestion of starch begins in the mouth.

★ **(M)** Saliva moistens the food.

As a result, the food is made slippery for easy swallowing.

3 ★ **(M)** Muscles of the **stomach** wall and **small intestine** mix food thoroughly with different juices containing digestive enzymes.

As a result, a liquid paste called **chyme** is formed.

As a result, food and digestive enzymes are brought into contact.

★ **(C)** Gastric juice, produced by **pits** in the stomach wall, contains **hydrochloric acid** and the enzymes **pepsin** and **rennin**.

Hydrochloric acid
• increases the acidity of the stomach contents.

As a result, bacteria in the food are killed.

As a result, the action of salivary amylase is stopped.

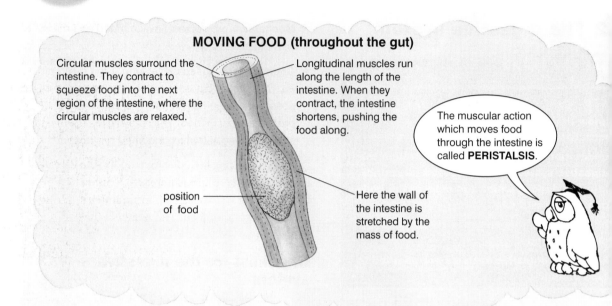

MOVING FOOD (throughout the gut)

Circular muscles surround the intestine. They contract to squeeze food into the next region of the intestine, where the circular muscles are relaxed.

Longitudinal muscles run along the length of the intestine. When they contract, the intestine shortens, pushing the food along.

The muscular action which moves food through the intestine is called **PERISTALSIS**.

position of food

Here the wall of the intestine is stretched by the mass of food.

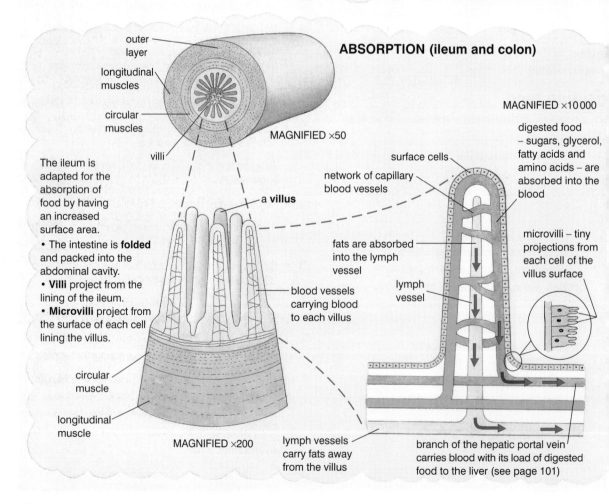

ABSORPTION (ileum and colon)

outer layer

longitudinal muscles

circular muscles

villi

MAGNIFIED ×50

The ileum is adapted for the absorption of food by having an increased surface area.
• The intestine is **folded** and packed into the abdominal cavity.
• **Villi** project from the lining of the ileum.
• **Microvilli** project from the surface of each cell lining the villus.

a **villus**

circular muscle

longitudinal muscle

MAGNIFIED ×200

surface cells

network of capillary blood vessels

fats are absorbed into the lymph vessel

blood vessels carrying blood to each villus

lymph vessel

lymph vessels carry fats away from the villus

MAGNIFIED ×10 000

digested food – sugars, glycerol, fatty acids and amino acids – are absorbed into the blood

microvilli – tiny projections from each cell of the villus surface

branch of the hepatic portal vein carries blood with its load of digested food to the liver (see page 101)

The digestive system – its structure and functions (checklist on pages 63 and 66)

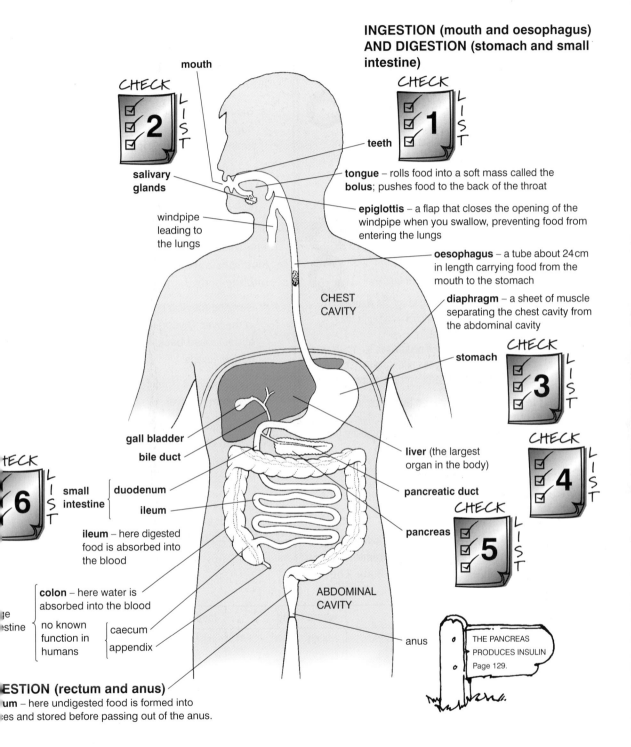

INGESTION (mouth and oesophagus) AND DIGESTION (stomach and small intestine)

CHECK LIST 2

mouth

CHECK LIST 1

teeth

salivary glands

tongue – rolls food into a soft mass called the **bolus**; pushes food to the back of the throat

windpipe leading to the lungs

epiglottis – a flap that closes the opening of the windpipe when you swallow, preventing food from entering the lungs

oesophagus – a tube about 24 cm in length carrying food from the mouth to the stomach

CHEST CAVITY

diaphragm – a sheet of muscle separating the chest cavity from the abdominal cavity

stomach

CHECK LIST 3

gall bladder

bile duct

liver (the largest organ in the body)

CHECK LIST 4

CHECK LIST 6

small intestine

duodenum

ileum

pancreatic duct

pancreas

CHECK LIST 5

ileum – here digested food is absorbed into the blood

colon – here water is absorbed into the blood

ge stine

no known function in humans

caecum

appendix

ABDOMINAL CAVITY

anus

THE PANCREAS PRODUCES INSULIN Page 129.

ESTION (rectum and anus)
um – here undigested food is formed into
es and stored before passing out of the anus.

Pepsin
- begins the digestion of protein.

Rennin
- clots milk, making it semi-solid.

As a result, milk stays in the gut long enough to be digested.

4 ★ **(C) Bile**, produced by the **liver**, is a green alkaline liquid which is stored in the gall bladder before release into the small intestine through the bile duct. It
- neutralises acid from the stomach
- breaks fat into small droplets (**emulsification**).

As a result, the surface area of fat exposed to the action of the enzyme **lipase** is increased.

As a result, fat is digested more quickly.

5 ★ **(C) Pancreatic juice**, produced by the **pancreas**, is released into the small intestine through the pancreatic duct. It contains
- **sodium carbonate** which neutralises stomach acid
- **carbohydrases**, **proteases** and **lipases** (see table below) which digest carbohydrate, protein and fat.

6 ★ **(C) Intestinal juice**, produced by glands in the wall of the **duodenum** and **ileum**, contains
- **carbohydrases** and **lipases** that complete the digestion of carbohydrates and fats.

Chemistry of digestion

Digestive enzymes catalyse the breakdown of food by hydrolysis.

Water splits large molecules of food which the body cannot absorb into smaller molecules which are suitable for absorption into the body. The table below summarises the process.

What happens to digested food?

Digested food is carried away from the ileum in the blood of the hepatic portal vein and in the fluid of the lymph vessels.

★ Blood transports water, sugars, glycerol and amino acids to the liver.

VEINS Page 99.
LYMPH VESSELS Page 102.

★ Lymph transports fats and fat-soluble vitamins to a vein in the neck where the substances enter the bloodstream.

The **liver** plays a major role in the metabolism food substances after they have been absorbed into the body.

★ Glucose may be converted to **glycogen** and stor in the liver. Glycogen may be hydrolysed to gluc and released back into the blood in response to body's needs.

★ Iron, obtained from destroyed red blood cells, is stored in the liver.

★ Amino acids in excess of the body's needs are broken down (a process called **deamination**) in the liver. Urea is poisonous and must be excreted in urine.

★ Amino acids are converted from one type into another in the liver (a process called **transamination**) according to the body's needs.

enzyme group	example	where found	food component	after digestion
carbohydrases (catalyse the digestion of carbohydrates)	amylase	mouth	starch	maltose
	maltase	small intestine	maltose	glucose
proteases (catalyse the digestion of proteins)	pepsin	stomach	protein	polypeptides
	chymotrypsin dipeptidase	small intestine	polypeptides dipeptides	dipeptides amino acids
lipases (catalyse the digestion of fats)	lipase	small intestine	fat	fatty acids + glycerol

Enzymes that digest carbohydrates, proteins and fats

3 Teeth and dentition

preview

t the end of this section you will:

know the structure of a tooth

understand the arrangement of teeth in the mouth

be able to identify adaptations of the skull and teeth to different diets.

diagram below shows the internal structure human tooth.

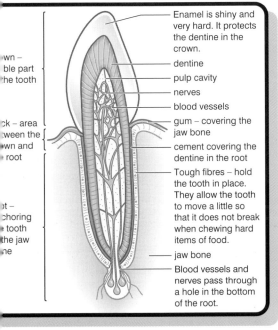

Enamel is shiny and very hard. It protects the dentine in the crown.

dentine

pulp cavity

nerves

blood vessels

gum – covering the jaw bone

cement covering the dentine in the root

Tough fibres – hold the tooth in place. They allow the tooth to move a little so that it does not break when chewing hard items of food.

jaw bone

Blood vessels and nerves pass through a hole in the bottom of the root.

wn – ble part he tooth

ck – area ween the wn and root

ot – choring e tooth the jaw ne

cture of a human tooth

pes of teeth

th are adapted to deal with food in different ys.

ncisors are chisel shaped for biting and cutting ood.

Canines are pointed for piercing, slashing and earing food.

Premolars and molars are large, with broad surfaces made uneven by bumps called cusps, for crushing and grinding food.

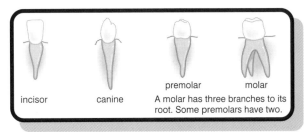

premolar   molar
A molar has three branches to its root. Some premolars have two.

incisor   canine

The four basic types of human teeth

The different types of teeth are positioned in the mouth according to their function. The word **dentition** is used to describe the number and arrangement of teeth in an animal.

Tooth enamel is the hardest substance in the body. It consists of calcium salts bound together by the protein **keratin**.

Omnivore dentition

Humans eat plants and meat (**omnivores**) and have all four basic types of teeth. The dentition is described in a **dental formula** using the letters shown here.

number and kind of teeth on each side of the upper jaw				
	2	1	2	3
i	c	p	m	
	2	1	2	3
number and kind of teeth on each side of the lower jaw				

Human dental formula

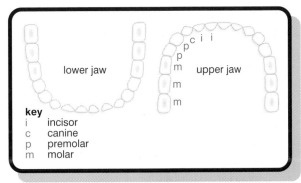

lower jaw   p c i i
              p
              p
              m   upper jaw
              m
              m

key
i   incisor
c   canine
p   premolar
m   molar

The arrangement of teeth in the adult human jaw

There are 32 teeth in total in the adult human jaw. In children there are 24 teeth, 20 of which have been gradually replaced by the permanent teeth by the age of about twelve. The teeth in children are called **milk teeth**. The third molars of the permanent teeth are called the **wisdom teeth** and do not appear until the age of about 20.

Herbivore dentition

The dentition of sheep and cattle is adapted to

* sweep grass into the mouth with the tongue poking through the **diastema** (gap between the canines and premolars)
* cut off the grass by the lower incisors nipping against the pad in the upper jaw
* grind the grass between the ridged, broad surfaces of the premolars and molars. The joint of the jaw moves from side to side as well as up and down which makes grinding food easier.

number and kind of teeth on each side of the upper jaw							
	0		0		3		3
i		c		p		m	
	3		1		3		3
number and kind of teeth on each side of the lower jaw							

Sheep dental formula

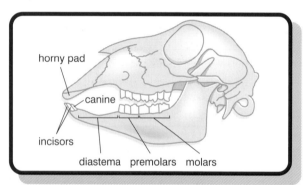

The arrangement of teeth in the jaws of a sheep. Notice that there are no canines in the upper jaw, and the canines in the lower jaw look like incisors.

Carnivore dentition

The dental formula of dogs and cats shows that their dentition is adapted to

* catch, hold and tear struggling prey with long, well developed canines
* cut through flesh and bone with incisors, premolars and molars.

Powerful jaw muscles ensure a firm grip and bite. The jaw joint only allows up-and-down movements.

number and kind of teeth on each side of the upper jaw							
	3		1		4		
i		c		p		m	
	3		1		4		
number and kind of teeth on each side of the lower jaw							

Dog dental formula

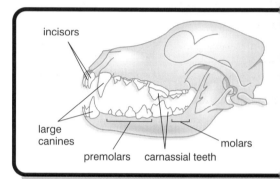

The arrangement of teeth in the jaws of a dog. Notice that the last premolar on each side of the upper jaw and the first molar on each of the lower jaw are fused to form the large **carnassial teeth**. They work like scissors, cutting through flesh and bone.

3.4 Reproduction

preview

At the end of this section you will:
* **be able to describe the main features of sperm and eggs**
* **understand the differences between externa fertilisation and internal fertilisation**
* **know about reproduction and development fish and mammals**
* **understand different strategies for the survival of young.**

Fact file

★ **Sperm** are the sex cells produced by males.

★ **Eggs** are the sex cells produced by females.

★ **Gametes** is a word that refers to both sperm and eggs.

e table compares sperm and eggs.

erm	eggs
• produced in large numbers	• produced in smaller numbers
• swim towards the egg	• do not move
• small	• large
• do not have food reserves	• have food reserves

rtilisation

•rtilisation occurs when sperm and egg join •ether. Sperm swim to the egg. Fertilisation •refore takes place in a liquid.

•ternal fertilisation occurs when organisms •ease sperm and eggs into the water in which •y live. Fertilisation of eggs takes place **outside** • body. For example, most types of fish •roduce by external fertilisation. The female •1 lays eggs in the water and the male fish •eases sperm near to the eggs. **Courtship** •aviour between male fish and female fish of • same species makes sure that, when released, •rm and eggs are close together.

Each fertilised egg contains a sac of **yolk**, which is a source of food for the developing embryo. It is surrounded and protected by a tough flexible membrane. A few days after fertilisation the young fish (called a **larva**) hatches from the egg. The yolk sac remains attached to the fish larva and provides food until the young fish is able to feed itself. The diagram shows the sequence.

Internal fertilisation occurs when the male releases sperm **inside** the female's body, where fertilisation of the egg(s) takes place. Sperm swim to the egg(s) in the liquid (called seminal fluid) which the male releases at the same time as the sperm.

Internal fertilisation is an adaptation that has evolved in insects, reptiles, birds and mammals for life on dry land. Seminal fluid provides the watery environment needed for sperm to swim to eggs inside the female body.

Survival of young

Broadly speaking, there is a relationship between the number of eggs produced and the amount of parental protection given to the eggs and to the young which develop after the eggs are fertilised. Fish and mammals illustrate two different strategies.

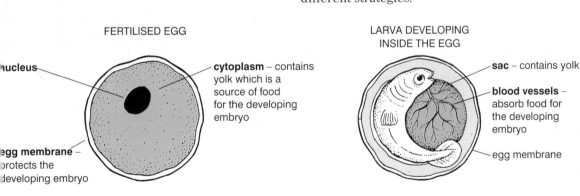

FERTILISED EGG

nucleus

cytoplasm – contains yolk which is a source of food for the developing embryo

egg membrane – protects the developing embryo

LARVA DEVELOPING INSIDE THE EGG

sac – contains yolk

blood vessels – absorb food for the developing embryo

egg membrane

LARVA HATCHED FROM THE EGG

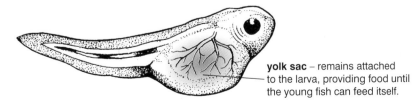

yolk sac – remains attached to the larva, providing food until the young fish can feed itself.

ges in the development of a fish from the fertilised egg (diagrams not to scale)

Fish: strategy 1

| Many eggs | → | The fertilised eggs or the young which develop from them receive little or no parental protection. | → | The chances of individual fertilised eggs or young surviving are **poor**. However, a few survive to become adults and reproduce the next generation. |

Fact file

The carp is a fish that lives in fresh water. Female carp each lay up to 800 000 eggs which the male fertilises.

Mammals: strategy 2

| Few eggs | → | The fertilised egg(s) and the embryo which develops from it receive food and a high level of protection inside the female body. | → | The female gives birth and the young depend on the parents for food and protection. The chances of the young surviving to become adults and reproduce the next generation are **good**. |

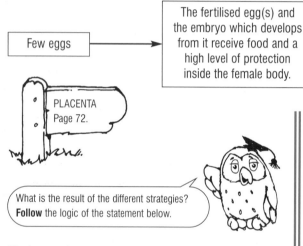

PLACENTA
Page 72.

What is the result of the different strategies?
Follow the logic of the statement below.

Statement

Fact 1: A pair of parents needs to produce only two fertilised eggs

Fact 2: which survive to become adults

Fact 3: that reproduce the next generation

Fact 4: for the numbers of a population to remain stable.

Remember the sequence of key facts:

two → survive → adults → reproduce → numbers stable

That's thinking – that's logic!

3.5 Having a baby

preview

At the end of this section you will:
- **know that physical and emotional changes take place as you grow up**
- **be able to identify the different parts of the human reproductive system**
- **understand how a baby develops inside its mother.**

Adolescence

The time line on page 71 shows the stages signposting human growth and development. T▌ onset of adolescence is marked by the development of **secondary sexual characteristi**▌ which are the physical features that help us tel▌ the difference between boys and girls. Differen▌ types of hormone help to develop and maintain the secondary sexual characteristics, which are summarised in the table.

Adolescence (including the teenage years) is a period of physical change. Feelings of sexual attraction to other people may also produce emotional turmoil. Many adolescents have sexu▌ feelings at some time in their growing up. Talki▌

out sexual feelings with someone you trust
en helps keep a sense of balance in
ationships.

The human reproductive system

A man and woman together produce a baby by
sexual reproduction. Their reproductive systems
are shown below.

Remember the sequence:

grows and develops grows and develops grows and develops

Baby ⟶ **Child** ⟶ **Adolescent** ⟶ **Adult**
0 years old   2–3 years old   11–14 years old   around
                              (includes the   20 years old
                              'teenage' years)

Time in years

---

oys	Girls
ubic hair develops around the penis and testes	Pubic hair develops around the opening to the vagina
enis becomes larger	Breasts develop and fat is laid down around the thighs
oice breaks and develops a deeper (lower) sound	Menstruation ('periods') starts, usually with the release of an egg from one of the ovaries every 28 days
air grows under the arms and on the chest, face and legs	Hair grows under the arms

---

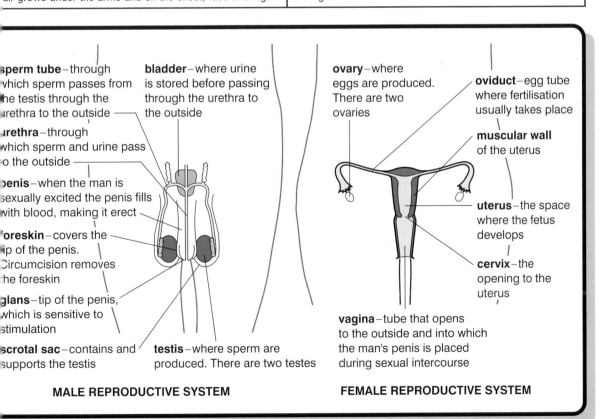

sperm tube – through which sperm passes from the testis through the urethra to the outside

urethra – through which sperm and urine pass to the outside

penis – when the man is sexually excited the penis fills with blood, making it erect

foreskin – covers the tip of the penis. Circumcision removes the foreskin

glans – tip of the penis, which is sensitive to stimulation

scrotal sac – contains and supports the testis

bladder – where urine is stored before passing through the urethra to the outside

testis – where sperm are produced. There are two testes

ovary – where eggs are produced. There are two ovaries

oviduct – egg tube where fertilisation usually takes place

muscular wall of the uterus

uterus – the space where the fetus develops

cervix – the opening to the uterus

vagina – tube that opens to the outside and into which the man's penis is placed during sexual intercourse

MALE REPRODUCTIVE SYSTEM

FEMALE REPRODUCTIVE SYSTEM

human reproductive system

During sexual intercourse the man's erect **penis** is placed inside the woman's **vagina**. Movement stimulates the penis and results in **ejaculation**. Sperm enter the **uterus** and swim along the tubes that connect the ovaries to the uterus. If an egg is present then one of the sperm may enter the egg and **fertilise** it. This is the moment of **conception**, and the woman is now **pregnant**. The figure opposite shows you what happens next.

The fetus develops *inside* the uterus. Oxygen, food and waste are exchanged between the mother and the fetus through the **placenta**. The **umbilical cord** attaches the fetus to the placenta.

SPERM and EGGS
Page 69.

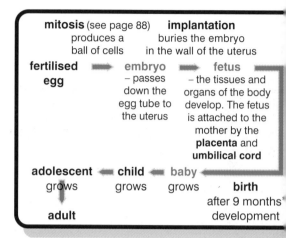

mitosis (see page 88)   implantation
        produces a   buries the embryo
        ball of cells   in the wall of the uterus

fertilised ➡ embryo ➡ fetus
    egg   – passes   – the tissues and
                down the   organs of the body
                egg tube to   develop. The fetus
                the uterus   is attached to the
                              mother by the
                              placenta and
                              umbilical cord

adolescent ⬅ child ⬅ baby ⬅
  grows   grows   grows   **birth**
                                   after 9 months'
  adult   development

Development of the fertilised egg

Being born

Birth usually occurs about 9 months after conception, when development is complete and the baby is fully grown. Babies born before 9 months are said to be **premature**. Providing birth doesn't happen too early, premature babies have a good chance of survival.

At birth the bag of watery liquid surrounding the baby bursts. Powerful contractions of the muscles of the wall of the uterus propel the baby, usually head first, through the vagina.

Being born is quite a shock, and stimulates the baby to start breathing. The placenta comes away from the uterus wall and passes out through the vagina as the **afterbirth**. The umbilical cord is clamped and cut near to the point where it joins the baby. Cutting the umbilical cord does not hurt the baby.

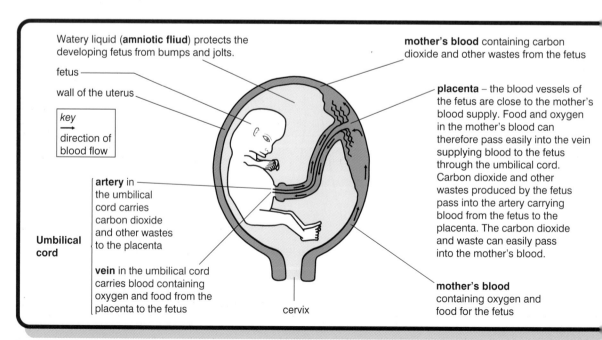

Watery liquid (**amniotic fliud**) protects the developing fetus from bumps and jolts.

fetus

wall of the uterus

key
➡
direction of blood flow

artery in the umbilical cord carries carbon dioxide and other wastes to the placenta

Umbilical cord

vein in the umbilical cord carries blood containing oxygen and food from the placenta to the fetus

cervix

mother's blood containing carbon dioxide and other wastes from the fetus

placenta – the blood vessels of the fetus are close to the mother's blood supply. Food and oxygen in the mother's blood can therefore pass easily into the vein supplying blood to the fetus through the umbilical cord. Carbon dioxide and other wastes produced by the fetus pass into the artery carrying blood from the fetus to the placenta. The carbon dioxide and waste can easily pass into the mother's blood.

mother's blood containing oxygen and food for the fetus

Caring for the fetus in the uterus

.6 Water and waste

preview

t the end of this section you will:

- be able to identify the ways in which the body gains and loses water
- know that urea is a waste product which is removed from the body in the urine
- be able to identify other waste products which are removed from the body
- know that the kidneys control the water content of the body and excrete wastes
- understand how antidiuretic hormone controls water balance
- be able to explain the implications of kidney disease and its treatment.

adult human needs to e in about 2500 cm³ of ter each day. The table low shows that the body ins and loses water rough its different tivities.

METABOLISM
Page 17.

tice that the daily gains and losses balance.

rea and other wastes

e body's metabolism produces waste stances.

★ **Respiration** produces carbon dioxide which is removed from the lungs.

★ **Deamination** in the cells of the liver breaks down amino acids in excess of the body's needs. The waste substance urea is produced.

RESPIRATION
Page 98.
LUNGS
Pages 98–99.

The process runs:

$$amino\ acids \rightarrow ammonia\ (very\ poisonous) + carbon\ dioxide \rightarrow urea\ (less\ poisonous)$$

Urea is carried in solution in the blood by way of the **renal arteries** to the kidneys where it is removed from the blood. The concept map on page 74 shows the kidney at work.

Fact file

★ Each kidney consists of about one million tiny tubules called **nephrons**.

★ The nephron is the working unit of the kidney. It controls the
  - concentration of salts in the body
  - water content of the body.

★ The nephrons are responsible for the excretion of urea and other wastes from the body.

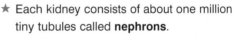

ains	volume (cm³/day)	losses	volume (cm³/day)
rinks: all drinks (tea, milk, coffee) are based n water	1400	Urine is a solution of wastes; mostly **urea**	1500
ood: fruit, vegetables and meat contain a lot f water	800	Sweat is a very dilute solution of salts that sweat glands release onto the body's surface, especially in hot weather. Evaporation of sweat cools the body	550
ell chemistry: chemical reactions in cells e.g. aerobic respiration) produce water	300	Breathing: breath exhaled from the lungs is saturated with water vapour	350
		Faeces is formed from the undigested remains of a meal. It contains water	100
	2500		2500

y gains/losses of water for an adult human

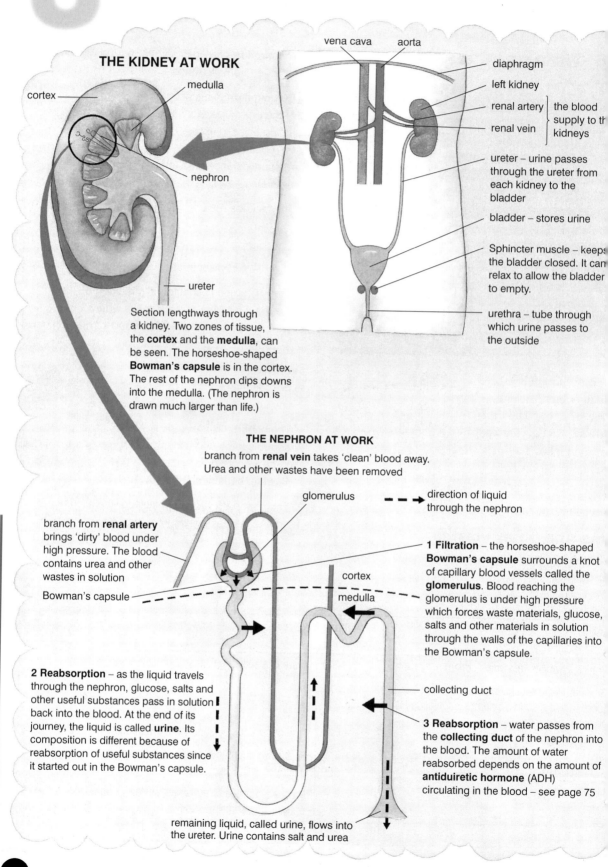

THE KIDNEY AT WORK

cortex

medulla

nephron

ureter

vena cava   aorta

diaphragm

left kidney

renal artery } the blood supply to the kidneys

renal vein }

ureter – urine passes through the ureter from each kidney to the bladder

bladder – stores urine

Sphincter muscle – keeps the bladder closed. It can relax to allow the bladder to empty.

urethra – tube through which urine passes to the outside

Section lengthways through a kidney. Two zones of tissue, the **cortex** and the **medulla**, can be seen. The horseshoe-shaped **Bowman's capsule** is in the cortex. The rest of the nephron dips downs into the medulla. (The nephron is drawn much larger than life.)

THE NEPHRON AT WORK

branch from **renal vein** takes 'clean' blood away. Urea and other wastes have been removed

glomerulus

- - → direction of liquid through the nephron

branch from **renal artery** brings 'dirty' blood under high pressure. The blood contains urea and other wastes in solution

Bowman's capsule

cortex

medulla

1 Filtration – the horseshoe-shaped **Bowman's capsule** surrounds a knot of capillary blood vessels called the **glomerulus**. Blood reaching the glomerulus is under high pressure which forces waste materials, glucose, salts and other materials in solution through the walls of the capillaries into the Bowman's capsule.

2 Reabsorption – as the liquid travels through the nephron, glucose, salts and other useful substances pass in solution back into the blood. At the end of its journey, the liquid is called **urine**. Its composition is different because of reabsorption of useful substances since it started out in the Bowman's capsule.

collecting duct

3 Reabsorption – water passes from the **collecting duct** of the nephron into the blood. The amount of water reabsorbed depends on the amount of **antiduiretic hormone** (ADH) circulating in the blood – see page 75

remaining liquid, called urine, flows into the ureter. Urine contains salt and urea

ontrolling water content

Diuresis is the flow of urine from the body.

Antidiuretic hormone counteracts diuresis.

As a result, the flow of urine from the body is reduced.

As a result, the loss of water from the body is reduced.

Ethanol (the alcohol in beer, wine and spirits) increases diuresis.

Antidiuretic hormone is one of the hormones produced by the pituitary gland at the base of the brain.

BRAIN
Page 105.

The diagram on page 00 shows how antidiuretic hormone controls the water content of the body.

Kidney disease

If a person's kidneys are not working (**kidney failure**) then

★ poisonous urea accumulates in the blood

★ water accumulates in the tissues of the body

As a result the person dies unless treated quickly.

BODY'S WATER CONTENT

Sensory receptors in the brain detect how much water is in the blood.

NOT ENOUGH WATER
Antidiuretic hormone (ADH) is produced from the pituitary gland.
As a result, the walls of the collecting duct of the nephron are more permeable ('leaky') to water.
As a result water is absorbed back into the body.

LOTS OF WATER
ced production of
uretic hormone (ADH) from
ituitary gland.
result, most of the surplus
is excreted through the
ys.

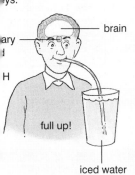

brain

ary

H

full up!

iced water

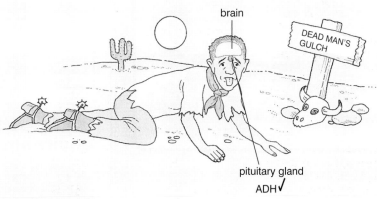

brain

DEAD MAN'S GULCH

pituitary gland
ADH ✔

kidney

ureter

water reabsorbed into the blood

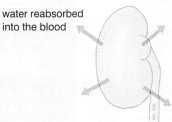

THE NEPHRON
Page 74.

ess water is excreted.
a result, large volumes of dilute urine are produced.

Very little water is excreted.
As a result, the urine is scanty and concentrated.

Kidney failure is usually treated by

★ dialysis. A **kidney machine** removes urea and other waste substances from the patient's blood. The diagram below shows you how the kidney machine works.

★ **transplant surgery**. A healthy kidney taken from a person (the donor) who has just died or from a living person (often a close relative) who wants to help the patient, is put inside the patient's body. The transplanted kidney is connected to the blood supply and to the bladder.

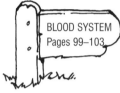

BLOOD SYSTEM
Pages 99–103

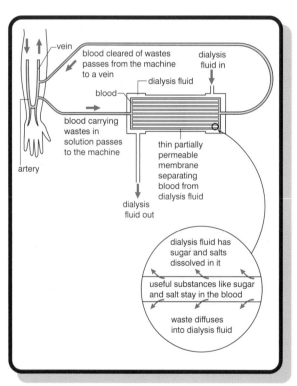

How a kidney machine works

The table summarises the benefits and limitatio of dialysis and transplant surgery.

dialysis	transplant surgery
Benefits The patient • does not have to take drugs for a lifetime	The patient • can enjoy a normal diet
• may be treated more quickly because the availability of kidney machines is restricted only by money	• does not depend on a kidney machine • has the freedom of a normal life
Limitations The patient • is restricted to eating a limited diet	The patient • is vulnerable to the body rejecting the new kidney. Transplanting a kidney from a close relative of the patient reduces the risk of rejection
• must always have easy access to a kidney machine	• must for their lifetime take drugs which reduce the risk of rejectic
• must be treated two or three times a week. Each treatment takes about 10 hours	• is vulnerable to infection: because the anti-rejectior drugs reduce resistance to disease-causing organisms (e.g. bacteria)
	• may have to wait for a suitable kidney because donor kidneys are in sho supply

3.7 Responding to the environment

preview

At the end of this section you will:

● **be able to give examples of changes in the environment which affect the responses of living things**

● **be able to describe the responses of woodlice to humidity and intensity of light**

● **understand the importance to living things of their responses to changes in the environmen**

e environment is changing all the time: mperature rises and falls; daytime alternates th night; dry periods are followed by rain.

anges in the environment are **stimuli**. They use living things to take action. The actions are lled **responses**.

Flowers close if temperature falls below a critical value.

Many types of animal sleep at night; many sleep during the day.

Tropical toads and frogs are very active when it rains after a long period of dry weather.

Birds reproduce during the spring when increasing length of day stimulates reproductive behaviour.

e change in the environment is the stimulus for e responses of organisms.

esponding to changes in the vironment

e diagram shows fferent types of European odlice. Why do they live der stones, in layers of tritus and other dark, mp places?

DETRITUS LAYER
Page 20.

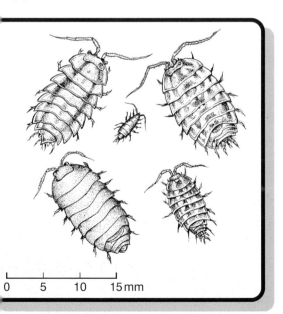

0   5   10   15 mm

opean woodlice. The scale shows the differences in size of the erent types (after J L Cloudsley-Thompson)

Woodlice quickly lose water from the body in dry air, threatening their survival.

As a result woodlice live in dark, damp places where the air is saturated with water vapour (**humid**).

Why? As a result loss of water from the body is reduced.

Look at *Moving Molecules* on page 85. **Remember** that substances diffuse from where the substance is in high concentration to where it is in low concentration. For woodlice the process runs:

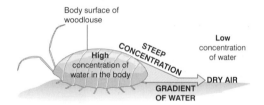

Body surface of woodlouse

High concentration of water in the body

STEEP CONCENTRATION

GRADIENT OF WATER

Low concentration of water

DRY AIR

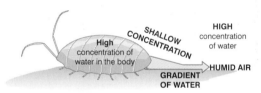

High concentration of water in the body

SHALLOW CONCENTRATION

GRADIENT OF WATER

HIGH concentration of water

HUMID AIR

How do woodlice find dark, damp places to live? Their **behaviour** makes sure they avoid environments where there would be a danger of them drying out.

★ The activity (walking) of woodlice depends on how **dry** the environment is (intensity of stimulus). The dryer it is the more active woodlice are, increasing their chances of finding damp environments. When they find a damp environment woodlice are much less active (stop moving).

★ Woodlice also respond to the intensity of **light**. The lighter it is, the more active woodlice are.

Why?

★ Light means lack of shelter.

As a result, the environment is more likely to be dry.

As a result woodlice are more likely to dry out and die.

You can use a **choice chamber** to investigate the responses of woodlice to dry/wet and light/dark environments. The diagram gives you the idea.

The choice chamber allows woodlice to choose different environment

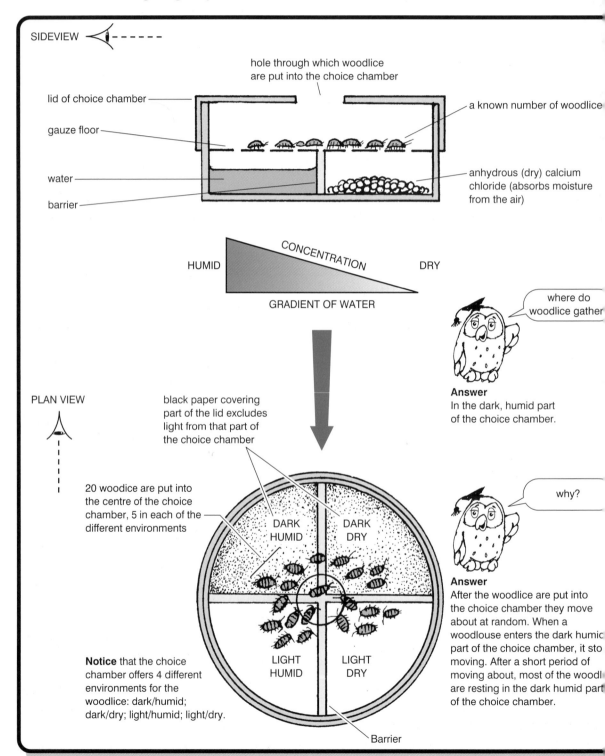

SIDEVIEW

hole through which woodlice are put into the choice chamber

lid of choice chamber

gauze floor

water

barrier

a known number of woodlice

anhydrous (dry) calcium chloride (absorbs moisture from the air)

CONCENTRATION

HUMID   DRY

GRADIENT OF WATER

where do woodlice gather

Answer
In the dark, humid part of the choice chamber.

PLAN VIEW

black paper covering part of the lid excludes light from that part of the choice chamber

20 woodice are put into the centre of the choice chamber, 5 in each of the different environments

DARK HUMID   DARK DRY

LIGHT HUMID   LIGHT DRY

why?

Answer
After the woodlice are put into the choice chamber they move about at random. When a woodlouse enters the dark humid part of the choice chamber, it sto moving. After a short period of moving about, most of the woodl are resting in the dark humid part of the choice chamber.

Notice that the choice chamber offers 4 different environments for the woodlice: dark/humid; dark/dry; light/humid; light/dry.

Barrier

rhythms of animal behaviour

odlice also have regular patterns of activity ythms) during the day and night (24 hours) and ween seasons.

ily rhythms

At night, woodlice do not respond to humidity as much as during daylight hours. When it is dark the environment is more humid and woodlice can leave their shelter with less risk of drying out.

As a result woodlice are able to explore new environments for food and mates.

As the intensity of light increases (early morning) woodlice return to their dark, damp shelters.

As a result the risk of drying out is reduced.

asonal rhythms

n spring, rain brings woodlice out of hibernation.

As a result, the **distribution** (where living things are found in the environment) of woodlice is affected, increasing their chances of finding new (damp!) environments in which to live.

dvantages of rhythmical activity

e intensity of light and humidity of the vironment are stimuli which trigger the rthmical activity of woodlice. Other **trigger nuli** which switch on different responses in ng things are listed on page 77.

Why are the responses important to the living things described?

Flowers close in response to a fall in temperature, avoiding possible frost damage.

Tropical toads and frogs are active when it rains, finding mates and reproducing. Their fertilised eggs and the **tadpoles** (young stages) which hatch from them, need water in which to develop into adults.

Birds reproduce in spring because the increase in day length coincides with an increase in the supply of food on which their young depend. The chances of the young surviving to become adults are improved.

Words to remember

You have read some important words in this chapter. Here's a list to remind you what the words in green mean.

Absorption of food	molecules of food pass through the wall of the intestine into the bloodstream
Baby	a fully developed fetus ready to be born
Dialysis	separation of substances in solution by using differences in the rate of diffusion of each substance through a membrane into another liquid
Diet	what we eat and drink
Embryo	the early stages of development after fertilisation. After a few weeks it develops into the fetus
Enzymes	catalysts made by living cells. Different enzymes help to speed up the digestion of food
Excretion	removal of wastes produced by the chemical reactions (metabolism) taking place in cells
Faeces	the semi-solid undigested remains of food from which water has been reabsorbed into the body from the large intestine
Fetus	the stage of development during which the tissues and organs of the body are formed
Hibernation	the inactive, dormant state of some animals (and plants) which helps them survive harsh conditions (e.g. cold, drought)
Hormones	chemicals produced and released by different tissues into the bloodstream
Hydrolysis	chemical breakdown of large molecules into smaller molecules by the addition of the elements of water
Metabolism	all of the chemical reactions which take place inside cells
Nutrients	substances essential for growth and other living processes

round-up

How much have you improved?
Work out your improvement index on page 136.

1 Simple tests identify the nutrients in different foods. Match the nutrient in column **A** with the test result that identifies the nutrient in column **B**.

A nutrients	B test results
starch	forms a milky emulsion when mixed with warm dilute ethanol
glucose	produces a violet/purple colour when mixed with dilute sodium hydroxide and a few drops of copper sulphate solution
fat	produces a blue/black colour when mixed with a few drops of iodine solution
protein	produces an orange colour when heated with Benedict's solution

[4]

2 Match each enzyme in column **A** with its role in digestion in column **B**.

A enzymes	B roles
amylase	digests maltose to glucose
pepsin	digests fat to fatty acids and glycerol
lipase	digests starch to maltose
maltase	digests protein to polypeptides

[4]

3 Briefly explain how antidiuretic hormone (ADH) keeps the water content of the body steady. [2]

4 Complete the following paragraph using the words below. Each word may be used once, more than once or not at all.

hardest   softest   keratin   calcium

Tooth enamel consists of _____ salts bound by the protein _____ . This enamel is the _____ substance in the body. [3

5 The diagram shows the reproductive system of a man. Name parts A–E.

E ─────
B ─────

[5

6 Leaves falling from trees form a thick layer of litter on the ground. Briefly state why you think that woodlice are often found in leaf litter. [4]

Well done if you've improved. Don't worry if you haven't. Take a break and try again.

nvestigating ells

How much do you already know?
Work out your score on page 136.

est yourself

Match each of the structures in column **A** with its function in column **B**.

A structures	B functions
cell membrane	fully permeable to substances in solution
chloroplast	partially permable to substances in solution
cell wall	contains the chromosomes
nucleus	where light energy is captured

[4]

Explain the difference between

a) a plasmolysed cell and a turgid cell [4]
b) a fully permeable membrane and a partially permeable membrane. [3]

What is a clone? [1]

a) Why do the cells of a tissue undergo mitosis?
b) In mitosis, what is the relationship between the number and type of chromosomes in the parent cell and in the daughter cells? [5]

Complete the following paragraph using the words below. Each word may be used once, more than once or not at all.

**types organism tissues organs cells
an organ**

Living things are made of _____ . Groups of similar _____ with similar functions form _____ that can work together as _____ .
A group of _____ working together form _____ system. [6]

Cellulose and chitin are important building materials in living things. Give an example of the use of each. [2]

The total magnification of a specimen seen under a light microscope is ×400. The magnifying power of the eyepiece lens is ×10. What is the magnifying power of the objective lens? [1]

The nutrients in food contain the elements carbon and oxygen. Which gas is released during aerobic respiration in cells? [1]

4.1 Looking at cells

preview

At the end of this section you will:
- **know that cells are the basic unit of living things**
- **understand how a light microscope works**
- **know that staining helps to improve contrast between different cell structures.**

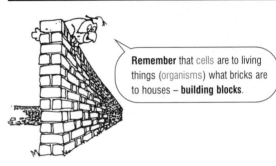

Remember that cells are to living things (organisms) what bricks are to houses – **building blocks**.

★ There are many different types of cell but all types of cell have **structures** that work in similar ways (**function**).

★ The structure of an organism depends on the way its cells are organised.

★ The function of an organism depends on the way its cells work.

Most cells are too small to be seen with the naked eye. The **light microscope** helps us see the structure of cells.

★ A lamp lights the specimen supported on a glass slide placed on the stage of the microscope.

★ The specimen is seen through two magnifying lenses.

Total magnification = Magnifying power of the eyepiece lens × Magnifying power of the objective lens

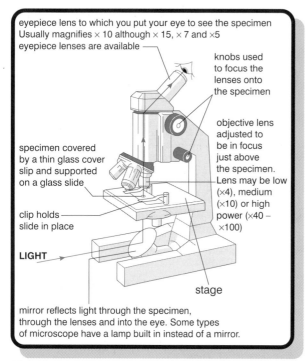

eyepiece lens to which you put your eye to see the specimen
Usually magnifies × 10 although × 15, × 7 and ×5
eyepiece lenses are available

knobs used
to focus the
lenses onto
the specimen

objective lens
adjusted to
be in focus
just above
the specimen.
Lens may be low
(×4), medium
(×10) or high
power (×40 –
×100)

specimen covered
by a thin glass cover
slip and supported
on a glass slide

clip holds
slide in place

LIGHT

stage

mirror reflects light through the specimen,
through the lenses and into the eye. Some types
of microscope have a lamp built in instead of a mirror.

A light microscope

Staining

The image of the cells seen through the
microscope is clear because:

★ The specimen is cut into very **thin** slices (ideally
only the thickness of a single layer of cells).

As a result light passes through the cells.

★ Cells are coloured (**stained**) with dyes (**stains**).
Some parts of a cell take up (absorb) stain better
than other parts (see table).

As a result the stained parts show up
against unstained areas.

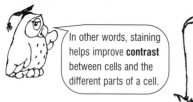

In other words, staining
helps improve **contrast**
between cells and the
different parts of a cell.

STRUCTURE
OF CELLS
Page 87.

stain	effect	cells
haematoxylin	stains nuclei blue	animal
eosin	stains cytoplasm pink	animal
iodine	stains nuclei brown	plant

Stains commonly used to pick out different cell structures

4.2 Cells at work

preview

**At the end of this section you will
know that:**

● plant cells and animal cells have structure
in common but are also different from one
another

● there are structures in cells which conver
energy from one form to another

● different types of cells are each specialise
to perform a particular biological task.

Cell functions

Remember that the structures that make up a
are organised in a way that depends on the
functions of the cell (the way it works).

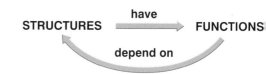

have

STRUCTURES ⟶ FUNCTIONS

depend on

The human body is made from more
than 200 different types of cell.

A plant is made from fewer different types of c

★ A group of similar cells makes a **tissue**.

★ Different tissues together make an **organ**.

★ Different organs combine to make an **organ syst**

How cells work

The diagram on pages 86–87 is the concept ma
for **cells at work**. The numbers on the diagram
refer to the checklist of points.

3 Investigating diffusion

preview

t the end of this section you will:

**understand that there is constant movement
of substances inside and into and out of cells**

know about diffusion and osmosis

**know that substances move along a
concentration gradient.**

ving molecules

ls need a non-stop supply of water and the
stances dissolved in it to stay alive. Substances
refore move inside and into and out of cells.

iffusion – the movement of a substance through
solution or gas *down* a concentration gradient
that is, from a region of high concentration to a
egion of low concentration)

smosis – the movement of **water** *down* a
oncentration gradient through a **partially
ermeable** membrane

t file   FACTS

Vhy does sugar sprinkled onto strawberries
urn pink? Because it makes the juice come out of
ie strawberries.

and out of cells

page 85 is the concept map for **Movement into
out of cells**. The numbers on the concept map
er to the checklist of points below.

ecklist for movement in
d out of cells   CHECK LIST

★ The molecules of a substance move at random, but
there is a better than even chance that some
molecules will spread from where they are highly
concentrated to where they are fewer in number.

As a result, there is a net movement of
molecules from where the substance is in high
concentration to where it is in low concentration.

★ Diffusion continues until the concentration of the
substance is the same throughout the gas or solution.

★ The greater the difference in concentration between
the regions, the steeper the concentration gradient and
the faster the substance diffuses.

---

2 ★ The flow of water through a membrane from a weak
solution to a more concentrated solution is called
osmosis.

★ A partially permeable membrane allows some
substances to pass through but stops others. The
passage of substances across such a membrane
depends on the
  • size of the molecules
  • size of the membrane
    pores
  • surface area of the cell
    membrane.

partially
permeable
membrane

★ The changes happening inside plant cells due to
osmosis bring about visible changes in the plant.
  • Plasmolysis causes a plant to **wilt** through lack of
    water.
  • The plant recovers following watering, which
    restores turgor to its cells.

★ Osmosis through a partially permeable membrane
continues until the concentrations of water on either
side of the membrane are equal.

Testing your understanding

Look again at the concept map for movement into
and out of cells on page 85.

★ The turgid plant cell is full of water. It does not burst
because it is surrounded by a tough cell wall.

★ In similar circumstances, an animal cell immersed
in a dilute solution (high concentration of water) of
a substance (e.g. sucrose) would:
  • *gain* water by osmosis
  • swell up
  • burst because there is no tough cell wall to
    protect the cell.

★ An animal cell immersed in a concentrated solution
(low concentration of water) of a substance (e.g.
sucrose) would:
  • *lose* water by osmosis
  • *shrink*
  • *crinkle*. The process is called **crenation**.

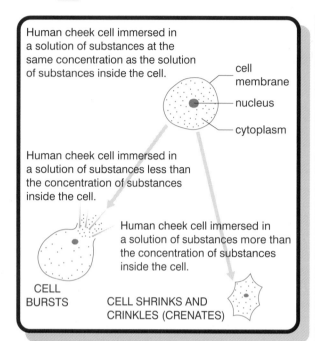

Human cheek cell immersed in a solution of substances at the same concentration as the solution of substances inside the cell.

cell membrane

nucleus

cytoplasm

Human cheek cell immersed in a solution of substances less than the concentration of substances inside the cell.

Human cheek cell immersed in a solution of substances more than the concentration of substances inside the cell.

CELL BURSTS

CELL SHRINKS AND CRINKLES (CRENATES)

Movement of water into and out of animal cells

Organ systems specialised for exchanging materials

We all exchange gases, food and other materials between our body and the environment. The exchange happens slowly by diffusion across body surfaces. Different organs and organ systems are specialised to increase the available surface area for the exchange of materials with their surroundings.

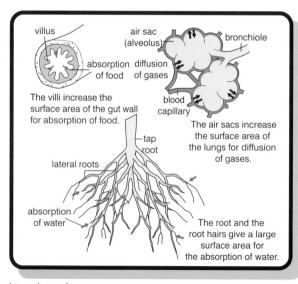

villus

air sac (alveolus)

bronchiole

absorption of food

diffusion of gases

The villi increase the surface area of the gut wall for absorption of food.

blood capillary

The air sacs increase the surface area of the lungs for diffusion of gases.

tap root

lateral roots

absorption of water

The root and the root hairs give a large surface area for the absorption of water.

Increasing surface area

4.4 Investigating cell division

preview

At the end of this section you will:
- **know that new cells (daughter cells) are formed when old cells (parent cells) divide into two**
- **understand that the cytoplasm and nucleus divide during cell division to produce identical cells.**

How cells divide

On page 89 is the concept map for division of the nucleus of a cell by mitosis. Study it carefully.

Remember

★ The nucleus of each **body cell** divides by **mitosis**.

★ The nucleus of each cell of the **sex organs** that gives rise to the sex cells (**gametes**) does so by type of division called **meiosis** (you do not need know the details). Sex cells are produced in the organs:
- the **testes** of the male and the **ovaries** of the female in mammals
- the **anthers** (male) and the **carpels** (female) in flowering plants.

SEXUAL REPRODUCT
Pages 71–72

Mitosis and meiosis

The nucleus contains **chromosomes**, each consisting of **deoxyribonucleic acid (DNA)** wou round a core of protein. In cell division, the chromosomes are passed from the **parent** cell t the new **daughter** cells. 'Daughter' does not mean that the cells are female. It means that they are the new cells formed as a result of cell division.

CHROMOSON
Page 115.

Mitosis produces daughter cells with the same number of chromosomes as the parent cell. The daughter cells are described as **diploid** (or **2n**).

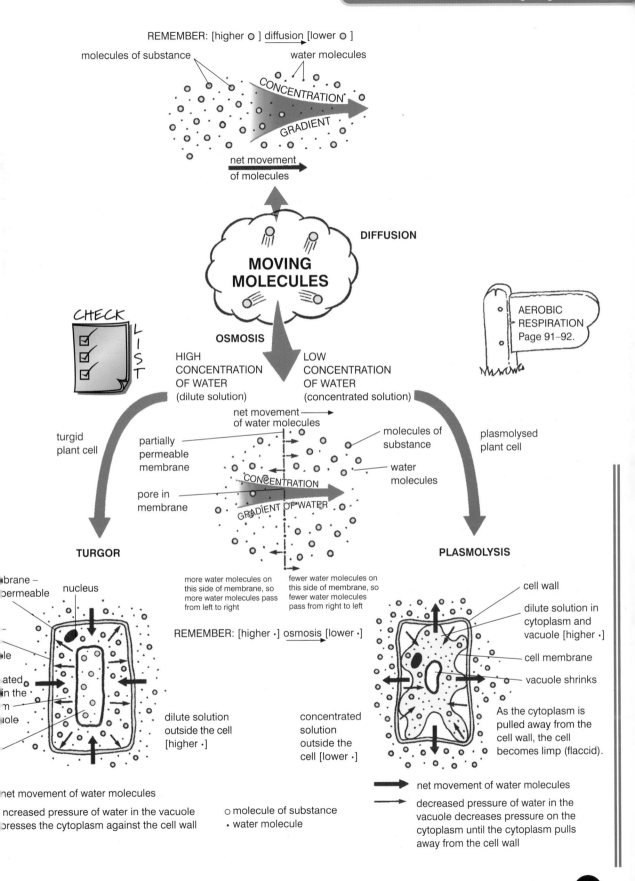

REMEMBER: [higher ○] diffusion [lower ○]

molecules of substance

water molecules

CONCENTRATION

GRADIENT

net movement
of molecules

**MOVING
MOLECULES**

DIFFUSION

CHECK
LIST

AEROBIC
RESPIRATION
Page 91–92.

OSMOSIS

HIGH
CONCENTRATION
OF WATER
(dilute solution)

LOW
CONCENTRATION
OF WATER
(concentrated solution)

turgid
plant cell

net movement
of water molecules

partially
permeable
membrane

molecules of
substance

water
molecules

CONCENTRATION

pore in
membrane

GRADIENT OF WATER

plasmolysed
plant cell

TURGOR

more water molecules on
this side of membrane, so
more water molecules pass
from left to right

fewer water molecules on
this side of membrane, so
fewer water molecules
pass from right to left

PLASMOLYSIS

cell wall

dilute solution in
cytoplasm and
vacuole [higher ·]

REMEMBER: [higher ·] osmosis [lower ·]

brane –
permeable

nucleus

le

ated
in the
m
ole

dilute solution
outside the cell
[higher ·]

concentrated
solution
outside the
cell [lower ·]

cell membrane

vacuole shrinks

As the cytoplasm is
pulled away from the
cell wall, the cell
becomes limp (flaccid).

net movement of water molecules

ncreased pressure of water in the vacuole
presses the cytoplasm against the cell wall

○ molecule of substance
· water molecule

net movement of water molecules

decreased pressure of water in the
vacuole decreases pressure on the
cytoplasm until the cytoplasm pulls
away from the cell wall

vement into and out of cells (checklist on page 83)

Checklist for cells at work

1 ★ During photosynthesis, oxygen is released into the environment.

★ During aerobic respiration, oxygen is used to release energy from food.

As a result, photosynthesis and aerobic respiration are stages in a cycle, the by-products of one forming the starting point of the other.

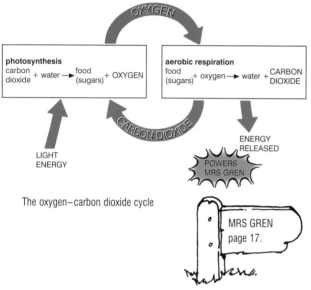

OXYGEN

| photosynthesis |
| carbon dioxide + water → food (sugars) + OXYGEN |

| aerobic respiration |
| food (sugars) + oxygen → water + CARBON DIOXIDE |

CARBON DIOXIDE

LIGHT ENERGY

ENERGY RELEASED

POWERS MRS GREN

The oxygen–carbon dioxide cycle

MRS GREN page 17.

2 ★ There are different types of cell for different functions.

★ Each type of cell is suited (**adapted**) for its function in the animal body or plant body.

★ A sheet of cells which covers a body surface is called an **epithelium**.

★ Red blood cells do not have nuclei.

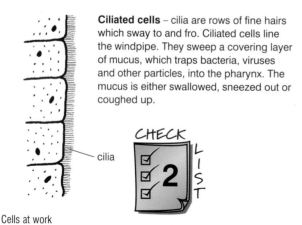

Ciliated cells – cilia are rows of fine hairs which sway to and fro. Ciliated cells line the windpipe. They sweep a covering layer of mucus, which traps bacteria, viruses and other particles, into the pharynx. The mucus is either swallowed, sneezed out or coughed up.

cilia

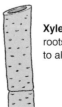

CHECK
LIST
2

Cells at work

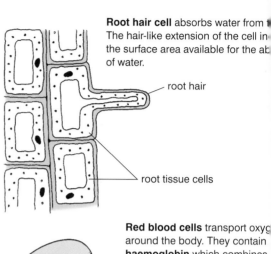

Root hair cell absorbs water from The hair-like extension of the cell in the surface area available for the ab of water.

root hair

root tissue cells

Red blood cells transport oxyg around the body. They contain **haemoglobin** which combines oxygen.

flattened disc shape increases surface area for the absorption of oxygen

VAF
OF (

sperm – the male sex cell which swims to the egg

tail-like flagellum lashes from side to side

ovum (egg) – the female sex ce is fertilised when a sperm fuses

The **leaf palisade cell** contains numerous chloroplasts.

chloroplasts

Xylem cells form tubes in the stem, roots and leaves, transporting water to all parts of the plant.

SUNLIGHT
ENERGY

chloroplast

PHOTO-
SYNTHESIS
Page 52–53.

captured by
chlorophyll

**PLANT CELLS
TRANSFORM
ENERGY**

CONVERTED TO CHEMICAL ENERGY
carbon + water → sugar + oxygen
dioxide (glucose)

CELL STRUCTURE

chloroplasts
contain chlorophyll
which captures light
energy

...ese are structures where
...ergy is released from the
...idation of glucose
...e structures are called
...tochondria).

nucleus contains the
chromosomes which
carry genes. It controls
the cell's activities,
including cell division

cell wall is
made of cellulose,
and is fully permeable
to substances in solution.
It gives the cell
shape and prevents the
cell from bursting

SUGAR

vacuole
contains cell sap
– a solution of
sugar and salts

cytoplasm is jelly-like
material which fills the
cell, giving it shape.
It is where chemical
reactions take place

PLANT CELL

1

...l **membrane** is
...tially permeable to
...bstances in solution

STRUCTURES FOUND IN
ANIMAL *AND* PLANT CELLS

STRUCTURES FOUND
ONLY IN PLANT CELLS

**ANIMAL AND
PLANT CELLS
TRANSFORM
ENERGY**

a mitochondrion

OXYGEN

MRS GREN
Page 17.

oxidation
of sugar
(glucose)

MOVEMENT
RESPIRATION
SENSITIVITY

POWERS

ENERGY

CHEMICAL ENERGY RELEASED
sugar + oxygen → carbon + water
(glucose) dioxide

AEROBIC
RESPIRATION
Page 91–92.

GROWTH
REPRODUCTION
EXCRETION
NUTRITION

Meiosis produces daughter cells with only half the number of chromosomes in the parent cell. The daughter cells are described as haploid (or *n*).

The importance of mitosis

The daughter cells each receive an identical full (diploid) set of chromosomes from the parent cell.

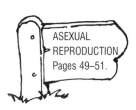

As a result, the parent cell and its daughter cells are genetically identical. They form a **clone**.

ASEXUAL REPRODUCTION Pages 49–51.

As a result, mitosis is the way in which living things

- **repair damage:** for example, mitosis replaces damaged skin cells with identical new skin cells
- **grow:** for example, the root of a plant grows because root tip cells divide by mitosis, increase in number and form new root tissue

DEVELOPMENT Page 48.

- **reproduce asexually:** for example, parts of stems can sprout roots and grow into new plants. The new individuals are genetically identical to the parents and are therefore clones.

The importance of meiosis

The daughter cells each receive a half (haploid) set of chromosomes from the parent cell.

As a result, during fertilisation (when sperm and egg join together), the chromosomes from each cell combine.

As a result, the fertilised egg (**zygote**) is diploid but inherits a new combination of genes contributed (50:50) from the parents.

As a result, the new individual inherits characteristics from both parents, not just from one parent as in asexual reproduction.

Make sure that you understand the table on page 51!

VARIATION Pages 112–113.

4.5 Investigating enzymes

preview

At the end of this section you will:

- **understand the meaning of the words** *catalyst, specific* **and** *optimum* **when talking about enzymes**
- **know that most enzymes are proteins**
- **be able to identify different reactions catalysed by enzymes**
- **know that temperature and pH affect the activity of enzymes.**

Catalysts help chemical reactions to take place but are not used up in the reaction itself. They are effective in *small* amounts, *unchanged* at the end of the reaction and usually *speed up* the rates of chemical reactions.

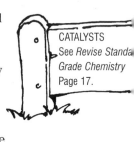

CATALYSTS See *Revise Standa Grade Chemistry* Page 17.

Enzymes are *organic* catalysts made by living cells. Most of them are proteins. They control t rates of chemical reactions in cells and speed the digestion of food in the gut. All of the featu of catalysts stated above are also features of enzymes. Also, enzymes are

★ **specific** in their action – each enzyme catalyses a particular chemical reaction or type of chemica reaction

★ sensitive to changes in **pH**

★ sensitive to changes in **temperature**.

The substance that the enzyme helps to react is called the **substrate**. The substance(s) formed i the reaction is called the **product(s)**. The featur of enzymes are shown in the diagram. Question and answers will help you to understand the diagram.

Dear student
Using just four chromosomes in the nucleus of the cell makes it easier for you to follow the process of mitosis. Notice the **before** and **after** features.
Luv
Owl

PARENT CELL

membrane
omosome
oplasm
lear
mbrane

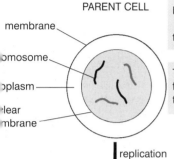

BEFORE MITOSIS
four chromosomes per cell:
the **diploid** number

The chromosomes shorten, fatten and become visible under the light microscope.

↓ replication

omatids
tromere

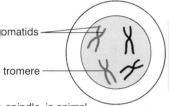

Each chromosome divides into a pair of identical (replica) **chromatids** joined to one another by the **centromere**.

spindle, in animal
ach pole forms
es which separate
division

ator of
cell

es
otein

Each pair of chromatids attaches to a spindle fibre by its centromere. The chromatids line up on the equator (middle) of the cell. The nuclear membrane has disappeared.

MITOSIS

of
tids

The chromatids are pulled apart by the spindle fibres and move to the opposite poles. The cell begins to divide.

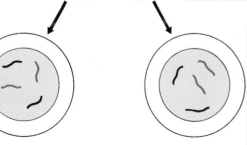

The chromatids are now the new chromosomes of the two daughter cells. A nuclear membrane forms around each group of chromosomes.

TWO DAUGHTER CELLS

division

AFTER MITOSIS
four chromosomes per cell: nucleus the **diploid** number

Plant cells

A thin slab-like structure called the **cell plate** extends outwards until it meets the sides of the cell. The cell plate divides the cytoplasm into two.

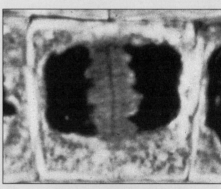

CELL DIVISION

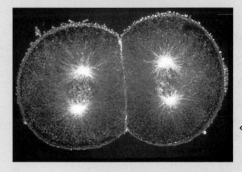

Animal cells

A furrow develops. It pinches the cell membrane in. As the furrow deepens the cell divides into two.

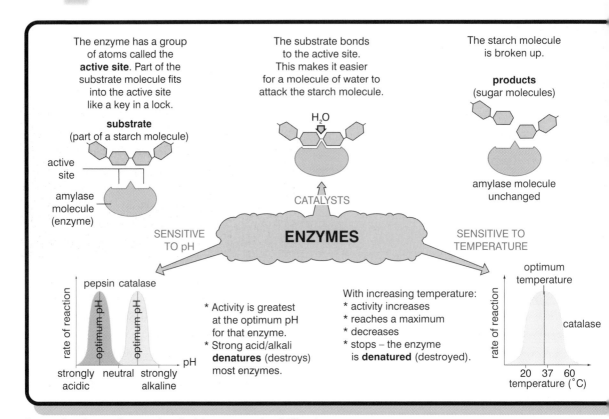

The enzyme has a group of atoms called the **active site**. Part of the substrate molecule fits into the active site like a key in a lock.

substrate (part of a starch molecule)

active site

amylase molecule (enzyme)

The substrate bonds to the active site. This makes it easier for a molecule of water to attack the starch molecule.

H_2O

CATALYSTS

The starch molecule is broken up.

products (sugar molecules)

amylase molecule unchanged

ENZYMES

SENSITIVE TO pH

SENSITIVE TO TEMPERATURE

pepsin catalase

rate of reaction / optimum pH / optimum pH / pH

strongly acidic   neutral   strongly alkaline

* Activity is greatest at the optimum pH for that enzyme.
* Strong acid/alkali **denatures** (destroys) most enzymes.

With increasing temperature:
* activity increases
* reaches a maximum
* decreases
* stops – the enzyme is **denatured** (destroyed).

optimum temperature

rate of reaction

catalase

20 37 60 temperature (°C)

Enzymes in action: *amylase*, *pepsin* and *catalase* are examples of different enzymes

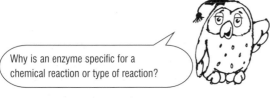

Why is an enzyme specific for a chemical reaction or type of reaction?

Answer
Only the shape of the particular substrate molecule (the 'key') fits the active site (the 'lock') of the enzyme.

Answer
The enzyme is less active and therefore less efficient as a catalyst.

Why, at pH/temperature values only slightly more or less than optimum, does the rate of reaction fall off?

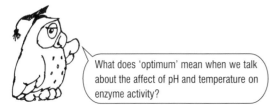

What does 'optimum' mean when we talk about the affect of pH and temperature on enzyme activity?

Answer
The 'optimum' is the value of pH or temperature at which the rate of chemical reaction is at a maximum.

Why is the enzyme less active and therefore less efficient?

Answer
A change of pH or temperature from optimum alters the shape of the active site of the enzyme preventing the substrate molecule from making a close fit. The enzyme-catalysed reaction may not occur.

hy enzymes?

e higher the temperature, the faster the rates
chemical reactions. However, temperatures
gher than 60°C kill most cells. Enzymes help
emical reactions in cells to occur quickly even
the relatively low temperatures (between 0°C
d 40°C) found in living things.

reaking and joining

fferent enzymes help break down large
olecules into smaller molecules. **Amylase** is an
ample of a 'breaking down' enzyme. The
agram on page 90 shows you how amylase
eaks down a large starch molecule into smaller
gar molecules.

psin and **catalase** are also examples of
eaking down' enzymes.

Pepsin breaks down large
protein molecules into
smaller peptide molecules.

PROTEIN
DIGESTION
Page 66.

Catalase breaks down
hydrogen peroxide into
water and oxygen. The reaction is very important
because hydrogen peroxide is a poisonous
substance produced by chemical reactions in cells.

her enzymes join up small molecules to make
ger molecules. **Phosphorylase** is an example of
joining up' enzyme. It joins small sugar
olecules together to make larger starch
olecules. The reaction is very important in
ants. Sugar made in the
aves by **photosynthesis** is
rned into starch, which
stored underground in
e roots (e.g. carrots) or
bers (e.g. potatoes).

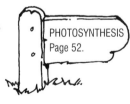

PHOTOSYNTHESIS
Page 52.

4.6 Investigating aerobic respiration

preview

At the end of this section you will:
- **be able to define aerobic respiration**
- **be able to explain why cells need energy**
- **know that the chemical reactions of aerobic respiration produce carbon dioxide and release heat**
- **be able to compare the energy content of different foods.**

The concept map on page 87 shows that during
aerobic respiration oxygen is used to release
energy from food. **Notice** that the chemical
reactions of aerobic respiration:

★ take place in the mitochondria of cells

★ release energy from the oxidation of food using
oxygen

★ produce carbon dioxide as a result of the oxidation
reactions.

Some of the energy released from food during
aerobic respiration is in the form of **heat**, which
helps warm the body. The remaining energy
powers life's activities. The mnemonic **MRS
GREN** on the concept map on page 87 helps you
to remember them. Don't forget to read the
signpost pointing you to page 17 for more details.

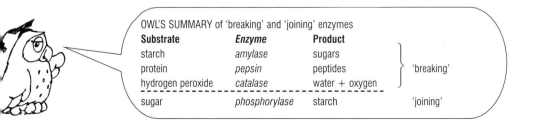

OWL'S SUMMARY of 'breaking' and 'joining' enzymes

Substrate	*Enzyme*	Product	
starch	*amylase*	sugars	
protein	*pepsin*	peptides	'breaking'
hydrogen peroxide	*catalase*	water + oxygen	
sugar	*phosphorylase*	starch	'joining'

Energy value of food

Oxygen is needed to burn food. Burning food in the controlled conditions of a laboratory investigation helps us to discover a value for the energy content of different foods.

★ The **energy value** of food is measured using an instrument called a **food calorimeter**, shown below.

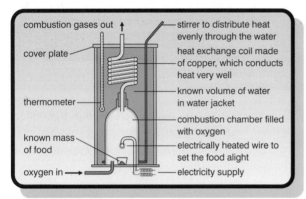

combustion gases out — stirrer to distribute heat evenly through the water

cover plate — heat exchange coil made of copper, which conducts heat very well

thermometer — known volume of water in water jacket

— combustion chamber filled with oxygen

known mass of food — electrically heated wire to set the food alight

oxygen in → — electricity supply

A food calorimeter

The burning food heats the surrounding water. The change in temperature of the water is used to work out the energy value of the food.

The energy released from food depends on the nutrients it contains:

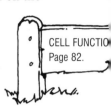

NUTRIENTS
Page 60.

- 17.2 kJ/g for carbohydrates
- 22.2 kJ/g for proteins
- 38.5 kJ/g for fats and oil.

Notice that fats and oils contain more chemical energy per gram than carbohydrates and proteins.

★ Although protein is an 'energy nutrient', its most important use in the body is for growth and repair.

Remember that the nutrients in food – carbohydrates, proteins, fats and oils – contain the **elements** *carbon* and *oxygen*. During aerobic respiration in cells the carbon and oxygen are released as *carbon dioxide* gas.

Why do cells need energy?

Remember that life's activities (**MRS GREN** – see page 17) depend on the functions of cells and that for cells to work properly they need energy. The energy comes from food and is released during aerobic respiration.

CELL FUNCTIO
Page 82.

Now you know the links between aerobic respiration, the release of energy from food, cell functions and life's activities.

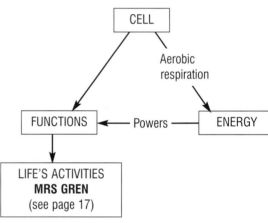

CELL

Aerobic respiration

FUNCTIONS ← Powers — ENERGY

LIFE'S ACTIVITIES
MRS GREN
(see page 17)

Cells need the energy released during aerobic respiration for:

★ growth

★ division

★ the contraction (shortening) of muscle cells

★ particular chemical reactions to take place.

The combination of sugar molecules to make starch (see page 61) is an example.

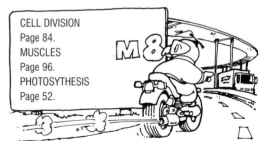

CELL DIVISION
Page 84.
MUSCLES
Page 96.
PHOTOSYTHESIS
Page 52.

Remember that **metabolism** is the word used to describe all of the chemical reactions which take place in cells (see page 17

Words to remember

u have read some important words in this
apter. Here's a list to remind you what the words
green mean.

apted	suited for a particular job or to a particular way of life in a particular environment
robic piration	uses oxygen to release energy from food
lls	microscopic structures containing cytoplasm surrounded by a membrane
ploid	daughter cells produced by division of a parent cell that each contain the same number of chromosomes as the parent cell
ploid	daughter cells produced by division of a parent cell so that each contain half

	the number of chromosomes of the parent cell
Meiosis	a type of cell division that produces haploid daughter cells. The cells of the sex organs that divide to produce gametes (sex cells) do so by meiosis
Mitosis	a type of cell division that produces diploid daughter cells. Body cells (other than the cells that give rise to gametes) divide by mitosis to repair damage, enable the body to grow and allow some organisms to reproduce asexually
Organism	any living thing
Proteins	large complex molecules, each made of many simpler amino acid molecules joined together
Specimen	a sample of material used for investigation or examination
Wilt	drooping of a plant through lack of water

round-up

How much have you improved?
Work out your improvement index on page 136.

1 Which of the structures listed below are found in
a) animal cells and plant cells **b)** plant cells only?

**nucleus   cell membrane   cell wall   large vacuole
mitochondria   chloroplasts   cytoplasm**   [7]

2 Describe what happens in the cells of a plant deprived
of water which is then watered. How will the
appearance of the plant change?   [5]

3 Complete the following paragraph using the words
below. Each word may be used once, more than once
or not at all.

**osmosis   faster   gains   down   slower   partially
energy   against**

The movement of a substance _____ a
concentration gradient is called diffusion. The steeper
the concentration gradient, the _____ is the rate
of diffusion.   [4]

4 Why is mitosis important for maintaining the health of
the tissues of the body?   [1]

5 The diagrams show a cell at different stages of
dividing by mitosis. Each stage is labelled with a
letter. Arrange the letters into the correct sequence of
stages of mitosis.   [6]

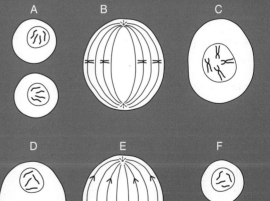

6 Name three substrates on which enzymes work and
the names of the products formed.   [3]

The body in action

How much do you already know?
Work out your score on page 137.

Making the link

The skeleton **supports** the body, and provides surfaces for the **attachment** of muscles. Nerve impulses stimulate muscles to contract. They pu on the skeleton, moving it. The sequence reads:

stimulate pull
nerve impulses → muscles contract → skeleton → MOV

NERVE IMPULSE
Page 103.

Test yourself

1 Explain the differences between
  a) bronchi and bronchioles   [2]
  b) lungs and alveoli   [2]
  c) aerobic respiration and anaerobic respiration   [4]
  d) breathing and gaseous exchange.   [3]

2 Explain how the heart functions as a double pump.   [4]

3 The different components of blood are listed in column **A**. Match each component with its correct description in column **B**.

A components	B descriptions
plasma	contain haemoglobin
red blood cells	contains dissolved food substances
white blood cells	produce antibodies [3]

4 The components of the reflex arc are listed as follows: sensory neurone, effector, relay neurone, receptor, motor neurone. Write the components in their correct order.   [4]

5 What is the function of each of these parts of the ear?
  a) the eardrum   **b)** the bones of the middle ear
  c) the pinna   **d)** the hair cells   [8]

5.1 Movement

preview

At the end of this section you will:

- **understand the principal parts of the human skeleton**
- **know the action of antagonistic muscles**
- **understand the structure of a joint and how it works**
- **be able to state the components of bone.**

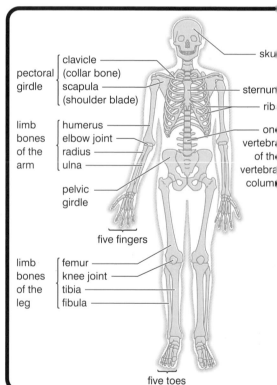

pectoral girdle
  clavicle (collar bone)
  scapula (shoulder blade)

limb bones of the arm
  humerus
  elbow joint
  radius
  ulna

pelvic girdle

five fingers

limb bones of the leg
  femur
  knee joint
  tibia
  fibula

sku

sternun
rib
on
vertebra
of th
vertebra
colum

five toes

The human skeleton

★ **Joints** are formed where the bones of the skeleton connect to one another.

★ **Ligaments** hold joints together.

★ **Tendons** attach muscles to the skeleton.

uscles contract (shorten) and relax
gthen), they move the bones at the joints.

parts of the human skeleton have different
tions.

tection

ull, which encloses and protects the brain.

ertebral column (backbone), made up of a series
bones called **vertebrae**. At the centre of each
ertebra is a channel called the **neural canal**,
hich forms a continuous space in the vertebral
olumn through which runs the **nerve cord**. The
ertebral column supports
e skull and the **pectoral**
nd **pelvic** girdles.

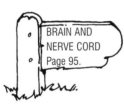

BRAIN AND
NERVE CORD
Page 95.

bs, which form a bony
age protecting the heart
nd lungs.

vement

elvic girdle, which links the legs with the vertebral
olumn. Its rigid framework allows forces on the
gs to be transmitted to the rest of the body.

ectoral girdle, which links the arms with the
ertebral column. Its flexibility gives the shoulders
nd arms freedom of movement.

mb bones, which in humans are long bones
inted at the elbows in the arms and at the knees
the legs. The limb bones also form joints with the
rdles at the hips and shoulders, and with the
ands at the wrists and the feet at the ankles. The
ints at the elbows, knees, wrists and ankles
nable the limbs to move freely.

pes of joint

re are different types of joint.

utures are fixed joints, for example the bones
f the skull.

Ball-and-socket joints are formed where the
upper long bones meet their respective
girdles. The flexibility of these joints allows
novement in *all* planes.

Hinge joints are formed at the elbow and
knee. They allow movement in *one* plane only.

The diagram shows a section through the joint
of an elbow. Notice that **synovial fluid** helps
reduce friction in the joint to a minimum. Ball-
and-socket joints and hinge joints contain
synovial fluid and are therefore synovial joints.

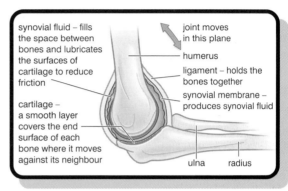

synovial fluid – fills the space between bones and lubricates the surfaces of cartilage to reduce friction

cartilage – a smooth layer covers the end surface of each bone where it moves against its neighbour

joint moves in this plane

humerus

ligament – holds the bones together

synovial membrane – produces synovial fluid

ulna   radius

The structure of the elbow joint

Bone and cartilage

Bone is a mixture of materials.

Cells called **osteoblasts** produce **collagen**, which
is a flexible fibrous material made of protein.

Calcium salts (mainly calcium phosphate) are
deposited in the collagen (protein), making it hard.

Blood vessels run through canals, called
Haversian canals, in the bone supplying the bone
tissue with oxygen and food substances. The
diagram gives you the idea.

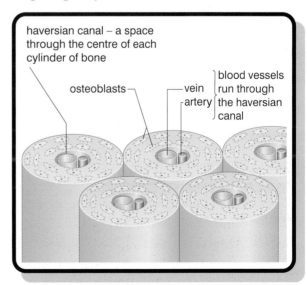

haversian canal – a space through the centre of each cylinder of bone

osteoblasts

vein
artery

blood vessels run through the haversian canal

The structure of bone

Cartilage is softer than bone. It contains fewer calcium salts. In humans, cartilage covers the ends of limb bones and helps reduce friction in the joints as bones move over one another.

Muscles in action

Contracting muscles pull on bones. A muscle will pull a bone in one direction; another muscle will pull the same bone in the opposite direction. In other words, muscles work in pairs, where one muscle of the pair has the opposite effect to its partner. We call these pairs **antagonistic pairs**.

The diagram shows you how the biceps and triceps raise (flex) and lower (extend) the arm.

The tendons, which attach muscles to bone, are **inelastic** (do not stretch), making precise, accurate movements possible.

Antagonistic pairs of muscles flap my wings!

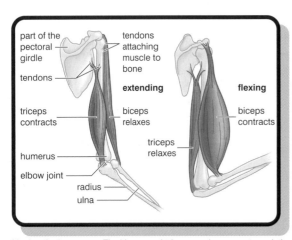

part of the pectoral girdle

tendons attaching muscle to bone

tendons

extending

flexing

triceps contracts

biceps relaxes

biceps contracts

triceps relaxes

humerus

elbow joint

radius

ulna

Moving the lower arm. The biceps and triceps work as an antagonistic pair of muscles.

5.2 The need for energy

preview

At the end of this section you will
- **know that breathing in (inhalation) takes a into the lungs and that breathing out (exhalation) pushes air out from the lungs**
- **be able to describe how breathing movements take place**
- **know that the exchange of gases (oxygen and carbon dioxide) happens in the lungs, between the air sacs (alveoli) and the capillary blood vessels**
- **understand that the blood transports oxyg to cells which use it for aerobic respiratio**

Fact file

★ People have different energy requirements depending on their
- **age** – on average young people have greate energy requirements than older people
- **gender** (male or female) – pregnancy and lactation (milk production) increase the energ requirements of women
- **activities** – any kind of activity increases a person's energy requirements.

FOOD ENERG
Page 92.

★ The rate at which the body uses energy is calle the **metabolic rate**. It is lowest (called the **basa metabolic rate**) when the body is at rest.

★ If a person eats more food than is necessary fo his/her energy needs, the excess is turned into

As a result, the person puts on weight.

★ To lose weight, a person can
- take more exercise, which increases energy ou
- eat less high-energy food, decreasing energ input.

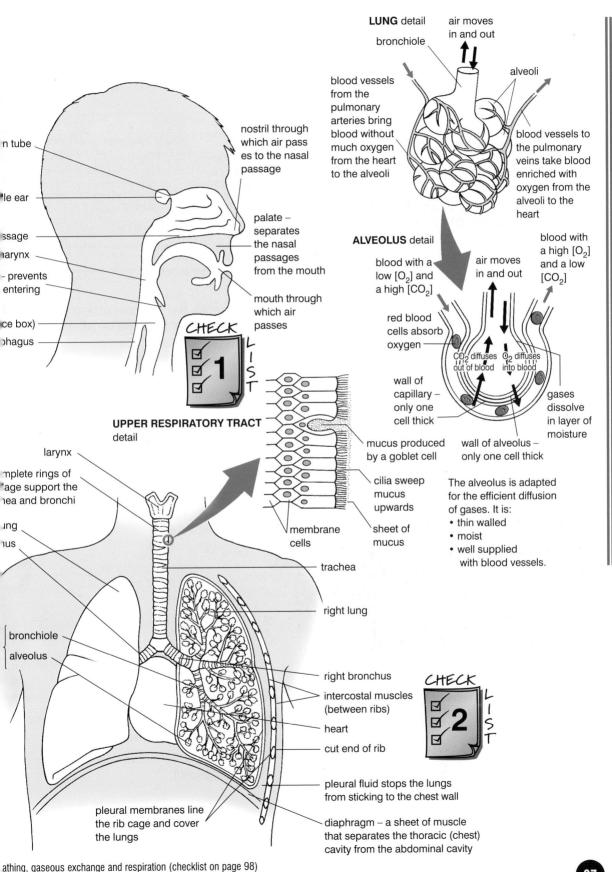

LUNG detail

air moves in and out

bronchiole

alveoli

blood vessels from the pulmonary arteries bring blood without much oxygen from the heart to the alveoli

blood vessels to the pulmonary veins take blood enriched with oxygen from the alveoli to the heart

ALVEOLUS detail

blood with a low [O$_2$] and a high [CO$_2$]

air moves in and out

blood with a high [O$_2$] and a low [CO$_2$]

red blood cells absorb oxygen

CO$_2$ diffuses out of blood   O$_2$ diffuses into blood

wall of capillary – only one cell thick

gases dissolve in layer of moisture

wall of alveolus – only one cell thick

The alveolus is adapted for the efficient diffusion of gases. It is:
• thin walled
• moist
• well supplied with blood vessels.

nostril through which air passes to the nasal passage

n tube

le ear

ssage

narynx

palate – separates the nasal passages from the mouth

- prevents entering

mouth through which air passes

ce box)

ohagus

CHECK
LIST
☑ ☑ ☑ **1**

UPPER RESPIRATORY TRACT detail

mucus produced by a goblet cell

cilia sweep mucus upwards

membrane cells

sheet of mucus

larynx

mplete rings of age support the nea and bronchi

ing

nus

trachea

right lung

bronchiole

alveolus

right bronchus

intercostal muscles (between ribs)

heart

cut end of rib

CHECK
LIST
☑ ☑ ☑ **2**

pleural fluid stops the lungs from sticking to the chest wall

pleural membranes line the rib cage and cover the lungs

diaphragm – a sheet of muscle that separates the thoracic (chest) cavity from the abdominal cavity

athing, gaseous exchange and respiration (checklist on page 98)

Respiration

Remember the distinction between respiration and gaseous exchange. Oxygen is used by cells to oxidise digested food substances (glucose) to release energy. The process is called **aerobic respiration**. The energy released from the **oxidation** of glucose powers the activities which define the characteristics of life (see page 17). The table shows that there is less oxygen in exhaled air than in inhaled air. This is because some of the oxygen is used by cells for aerobic respiration. There is more carbon dioxide in exhaled air than in inhaled air because carbon dioxide is produced by the chemical reactions of aerobic respiration.

AEROBIC RESPIRATION Pages 91–92.

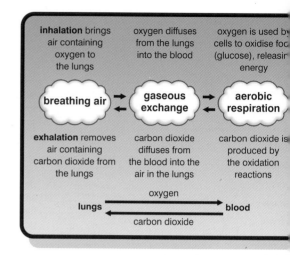

gas	amount in inhaled air / %	amount in exhaled air / %
nitrogen	78	78
oxygen	21	16
noble gases	1	1
carbon dioxide	0.03	4
water vapour	0	1

Differences between inhaled and exhaled air

★ How does oxygen reach cells?

★ How does carbon dioxide leave cells?

The answer is by gaseous exchange.

Now you can see the link between breathing air, gaseous exchange and aerobic respiration.

On page 97 is the concept map for **breathing, gaseous exchange and respiration**. The numbers on the concept map refer to the checklist below.

Checklist for breathing, gaseous exchange and respiration

CHECK

1 ★ The **upper respiratory tract** is a tube from the nose and mouth to the lungs.
  • It is well supplied with blood.

  As a result, inhaled air is warmed to body temperature.

  • Hairs in the nasal passage filter out large dust particles.
  • The lining of **mucus** traps bacteria, viruses and dust particles.
  • Hair-like **cilia** sweep the mucus into the **pharynx** where it is either swallowed, sneezed out or coughed up.

  As a result, the air entering the lungs is cleaned and freed of disease-causing microorganisms.

2 ★ The network of **bronchioles** in the lungs form the **bronchial tree**.

  ★ The millions of **alveoli** in a pair of human lungs form a surface area of about 90 m^2.

  As a result, gaseous exchange is very efficient.

3 ★ Blood with a low concentration of oxygen ($[O_2]$) is said to be **deoxygenated**.

  ★ Blood with a high concentration of oxygen is said to be **oxygenated**.

eathing movements

ribs and **diaphragm** form an elastic cage
und the lungs. As they move, the pressure in
lungs changes. This change in pressure
ses **inhaling** (breathing in) and **exhaling**
athing out).

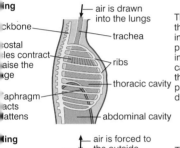

ing

- air is drawn into the lungs
ckbone
ostal
les contract
aise the
ge — trachea
— ribs
— thoracic cavity
aphragm
acts
lattens
— abdominal cavity

The volume of the thoracic cavity increases. The pressure of air inside the thoracic cavity becomes less than atmospheric pressure, so air is drawn into the lungs.

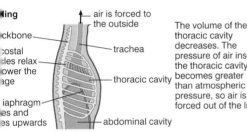

ling

- air is forced to the outside
ckbone
ostal
les relax
ower the
ge — trachea
— thoracic cavity
iaphragm
es and
es upwards
— abdominal cavity

The volume of the thoracic cavity decreases. The pressure of air inside the thoracic cavity becomes greater than atmospheric pressure, so air is forced out of the lungs.

ng and exhaling

3 Blood and the rculatory system

preview

t the end of this section you will:

be able to identify the different components of blood

understand the functions of blood

understand why the heart is a double pump

know that capillary blood vessels link arteries and veins.

t file

The **heart** is a pump.

Blood is a liquid containing different cells.

Arteries, veins and **capillaries** are tube-like vessels through which blood flows.

$$\text{eart} \xrightarrow{\text{pumps}} \text{blood} \xrightarrow{\text{through}} \text{blood vessels}$$

Moving blood around

The circulatory system consists of tubes (arteries, veins and capillaries) through which blood is pumped by the heart. Blood carries oxygen, digested food, hormones and other substances *to* the tissues and organs of the body that need them. Blood also carries carbon dioxide and other waste substances produced by the metabolism of cells *from* the tissues and organs of the body. On page 101 is the concept map for **blood and the circulatory system**. The numbers on the concept map refer to the checklist below.

Checklist for blood and the circulatory system

1 ★ **Red blood cells** are made in the **marrow** of the limb bones, ribs and vertebrae.

LIVER
Page 66.

★ Old red blood cells are destroyed in the liver.

★ Red blood cells contain haemoglobin, which combines with oxygen.

★ **White blood cells** originate in the **bone marrow** and **spleen**. They help defend the body against diseases.

2 ★ **Arteries** carry blood *from* the heart.

Veins carry blood *to* the heart.

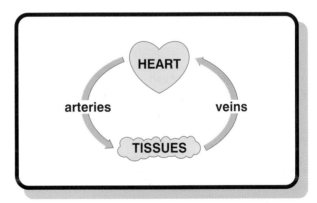

3 ★ Heart (**cardiac**) muscle contracts and relaxes rhythmically for a lifetime.

★ **Valves** direct the flow of blood through the heart.

★ The heartbeat is a two-tone sound. Contraction of the auricles (atria) forces blood into the ventricles. Contraction of the ventricles forces blood into the

BLOOD

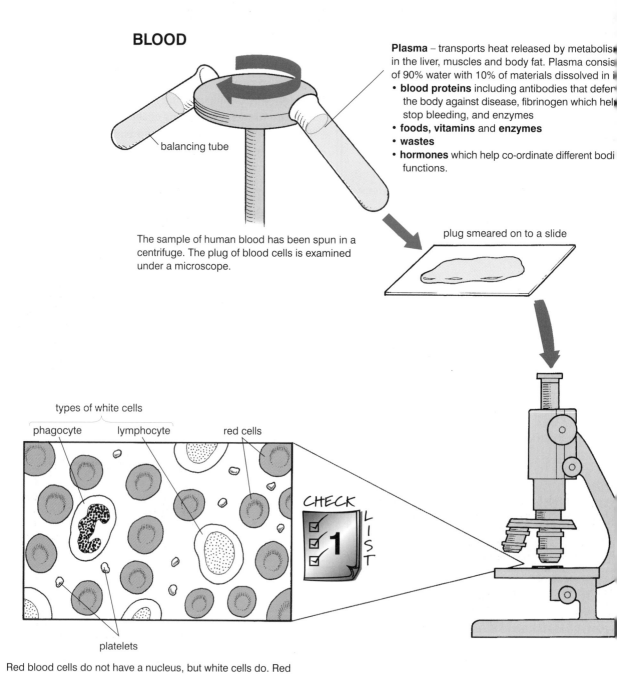

balancing tube

The sample of human blood has been spun in a centrifuge. The plug of blood cells is examined under a microscope.

Plasma – transports heat released by metabolis in the liver, muscles and body fat. Plasma consis of 90% water with 10% of materials dissolved in i
- **blood proteins** including antibodies that defer the body against disease, fibrinogen which hel stop bleeding, and enzymes
- **foods, vitamins** and **enzymes**
- **wastes**
- **hormones** which help co-ordinate different bodi functions.

plug smeared on to a slide

types of white cells

phagocyte   lymphocyte   red cells

CHECK
☑
☑ **1**
☑

LIST

platelets

Red blood cells do not have a nucleus, but white cells do. Red cells are packed with the pigment haemoglobin which gives cells their red colour. Notice the characteristic shapes of the nuclei of phagocytes and lymphocytes. Platelets look like fragments of red cells.

Blood and the circulatory system (checklist on pages 99 and 102)

BLOOD SYSTEM

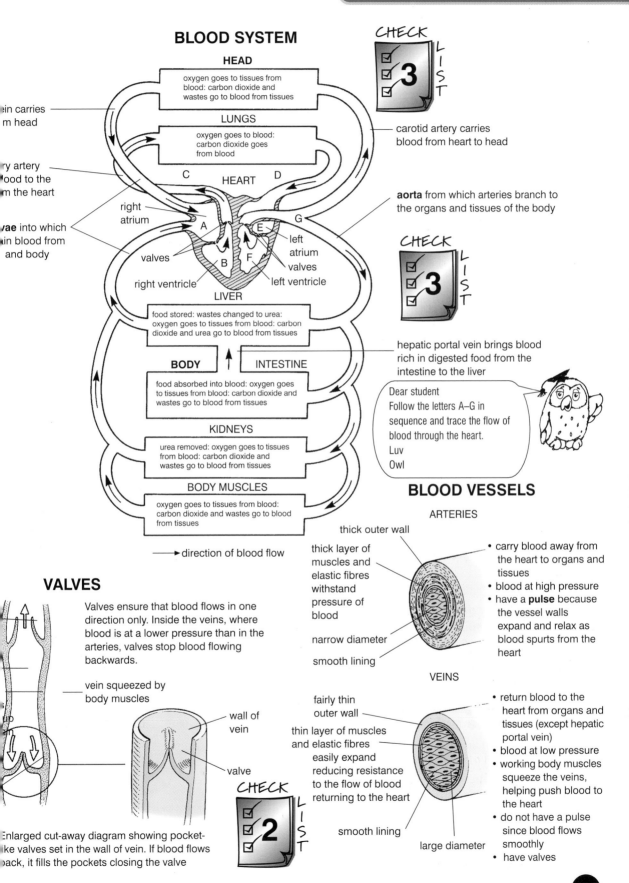

HEAD
oxygen goes to tissues from blood: carbon dioxide and wastes go to blood from tissues

ein carries m head

ry artery ood to the m the heart

vae into which in blood from and body

LUNGS
oxygen goes to blood: carbon dioxide goes from blood

C   D

HEART

right atrium

A   G

valves

B   F

right ventricle

E
left atrium
valves
left ventricle

LIVER
food stored: wastes changed to urea: oxygen goes to tissues from blood: carbon dioxide and urea go to blood from tissues

BODY   ↑   **INTESTINE**

food absorbed into blood: oxygen goes to tissues from blood: carbon dioxide and wastes go to blood from tissues

KIDNEYS
urea removed: oxygen goes to tissues from blood: carbon dioxide and wastes go to blood from tissues

BODY MUSCLES
oxygen goes to tissues from blood: carbon dioxide and wastes go to blood from tissues

→ direction of blood flow

carotid artery carries blood from heart to head

aorta from which arteries branch to the organs and tissues of the body

hepatic portal vein brings blood rich in digested food from the intestine to the liver

Dear student
Follow the letters A–G in sequence and trace the flow of blood through the heart.
Luv
Owl

CHECK LIST 3

CHECK LIST 3

BLOOD VESSELS

ARTERIES

thick outer wall

thick layer of muscles and elastic fibres withstand pressure of blood

narrow diameter

smooth lining

- carry blood away from the heart to organs and tissues
- blood at high pressure
- have a **pulse** because the vessel walls expand and relax as blood spurts from the heart

VEINS

fairly thin outer wall

thin layer of muscles and elastic fibres easily expand reducing resistance to the flow of blood returning to the heart

smooth lining

large diameter

- return blood to the heart from organs and tissues (except hepatic portal vein)
- blood at low pressure
- working body muscles squeeze the veins, helping push blood to the heart
- do not have a pulse since blood flows smoothly
- have valves

VALVES

Valves ensure that blood flows in one direction only. Inside the veins, where blood is at a lower pressure than in the arteries, valves stop blood flowing backwards.

vein squeezed by body muscles

wall of vein

valve

Enlarged cut-away diagram showing pocket-ke valves set in the wall of vein. If blood flows ack, it fills the pockets closing the valve

CHECK LIST 2

pulmonary artery (from the right ventricle) and aorta (from the left ventricle)

★ The heart is a **double pump**.

★ The beating of the heart is controlled by a pacemaker.

  As a result, the heart beats on average 72 times a minute.

★ The coronary arteries supply blood to the surface of the heart.

  As a result the heart muscle receives the supplies of food and oxygen carried by the blood.

4 ★ The arteries and veins in the human body form two circuits:
  • the lung circuit
  • the head and body circuit.

★ The wall of the left ventricle is thicker than the wall of the right ventricle.

  As a result its contractions are powerful enough to pump blood around the head and body circuit. The distance is greater than the distance the right ventricle pumps blood around the lung circuit.

Testing your understanding

Examinations test your understanding of ideas and important principles. Be sure you have grasped the arrangement of the **hepatic portal vein** and the role of the **pulmonary artery** and **pulmonary vein**.

Understanding the hepatic portal vein

Veins carry blood to the heart. The **hepatic portal vein** is the exception. Notice on page 101 that it carries blood with its load of digested food from the intestine to the liver.

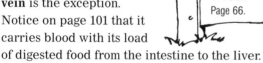
ABSORPTION OF FOOD Page 66.

Understanding the pulmonary artery and the pulmonary vein

Arteries are often described as carriers of oxygenated blood (often coloured red on diagrams), and veins as carriers of deoxygenated blood (often coloured blue on diagrams). However, the **pulmonary artery** carries deoxygenated blood from the heart to the lungs. The **pulmonary vein** carries oxygenated blood from the lungs to the heart.

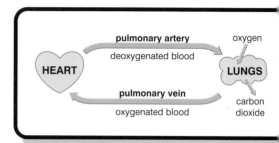

Remember that the haemoglobin in red blood cells combines with oxygen in tissues where the concentration of oxygen is high. **Oxyhaemoglobin** is formed. It breaks down in tissues where the concentration of oxygen is lo releasing oxygen to the tissues.

$$\text{Haemoglobin} + \text{oxygen} \underset{\text{other body tissues}}{\overset{\text{lung tissue}}{\rightleftarrows}} \text{oxyhaemoglo}$$

Capillaries – fact file

Capillaries are tiny blood vessels, 0.001 mm in diameter.

★ The walls of capillary blood vessels are one cell thick.

  As a result, substances easily diffuse between blood in the capillaries and the surrounding tissues.

★ Arteries branch into arterioles, which divide into capillaries. The capillaries form dense networks called **beds** in the tissues of the body, providing

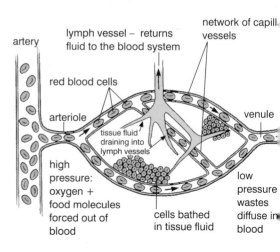

Capillaries at work

large surface area for the efficient exchange of materials between the blood and tissues.

The blood in capillaries supplies nearby cells with oxygen, food molecules and other substances. It also carries away carbon dioxide and other wastes produced by the cells' metabolism.

Tissue fluid carries oxygen, food and other substances to the cells. This fluid is blood plasma that has been forced out through the thin capillary walls by the pressure of the blood inside.

Red blood cells squeeze through the smallest capillaries in single file.

As a result, the pressure drops as blood passes through the capillaries from the artery to the vein.

.4 Co-ordination and the nervous system

preview

t the end of this section you will know that:

- stimuli are converted by receptors into signals called nerve impulses, to which the body can respond
- neurones (nerve cells) conduct nerve impulses to muscles, which respond by contracting
- muscles are called effectors
- nerves are formed from bundles of neurones and are the link between stimulus and response.

he process runs:

mulus → receptor → nerves → effector → response

timulus and response

stimulus is a change in the environment which uses a living organism to take action. A sponse is the action that the living organism kes. The **nervous system** links stimuli and sponses, co-ordinating the body's physical tivities. This is the sequence of events:

- ★ **Sensory receptor cells** detect stimuli and convert them into **nerve impulses**, to which the body can respond.

- ★ Nerve impulses are minute electrical disturbances.

- ★ **Neurones** (nerve cells) conduct nerve impulses to **effectors** (muscles or glands). Muscles respond to nerve impulses by contracting; glands respond by secreting substances. For example, the adrenal glands respond to nerve impulses by producing the hormone adrenaline, which helps the body cope with sudden stress.

The nervous system

On pages 105–106 is the concept map for **the nervous system**. The numbers on the concept map refer to the checklist.

Checklist for the nervous system

1 ★ Each **nerve** of the nervous system consists of a bundle of **neurones**.

★ Each neurone transmits nerve impulses to an **effector** (muscle or gland).

★ Nerve impulses are minute electrical disturbances which carry information about stimuli.

★ Nerve impulses stimulate effectors to respond to stimuli in a useful way.

★ A nerve impulse takes just milliseconds to travel along a neurone.

2 ★ Reflex responses happen before the brain has had time to process the nerve impulses carrying the information about the stimulus.

★ When the brain catches up with events, it brings about the next set of reactions – such as a shout of pain.

★ **Ascending fibres** carry nerve impulses to the brain.

As a result, the brain receives information about the stimulus causing the reflex response.

★ Nerve impulses from the brain are transmitted to effectors.

As a result, the reflex response is brought under conscious control.

3 ★ The human brain weighs approximately 1.3 kg and occupies a volume of about 1500 cm³.

★ Around 6 million neurones make up 1 cm³ of brain matter.

★ Memory and learning are under the brain's control.

★ Different drugs affect the brain. For example, ethanol (the alcohol in beers, wines and spirits) depresses the activity of the cerebral cortex, affecting judgement and the control of movement.

5.5 Detecting stimuli

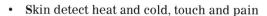

preview

At the end of this section you will know that:

● **the sense organs consist of sensory cells which are adapted to detect a particular type of stimulus.**

Handy hint

The sensory cells of the

Hints & Tips

- **S**kin detect heat and cold, touch and pain
- **N**ose detect chemicals
- **E**ye detect light
- **E**ar detect sound
- **T**ongue detect chemicals.

Thinking of the mnemonic **SNEET** will help you remember the major sense organs of the body.

Sensing the surroundings

On page 107 is the concept map for eyes and ears. The numbers refer to the checklist below.

Checklist for eyes and ears

CHECK LIST

1 ★ **Tears** lubricate the surface of the eye. They contain the enzyme **lysozyme** which kills bacteria.

★ The **iris** of the eye is usually coloured brown, blue or green.

★ A pair of human eyes contains around 130 million **rods** and 7 million **cones**.

★ **Cone** cells are packed most densely in the region of the **fovea** and respond to bright light.

As a result, objects are seen most clearly if looked at straight on.

★ **Rod** cells occur mostly near the edges of the retina and respond to dim light.

As a result, objects are seen less clearly out of the corner of the eye.

2 ★ Loudness is measured in **decibels**. The faintest sou that the ear can hear is defined as zero decibels.

★ The response of the ear to different levels of loudne varies with frequency. The ear is most sensitive to frequencies around 3000 Hz, and can detect the softest sounds. It is completely insensitive to soun over 18 000 Hz and cannot detect them.

★ The walls of the ear tube produce wax, which keep the eardrum soft and supple.

★ Two ears are better than one for judging the *directi* of sound.

Check the vibrations

In the concept map, notice the different structu in the ear vibrating in response to sound waves striking the eardrum. The sequence reads:

- eardrum
- bones of the middle ear
- oval window
- fluid in the cochlea
- basilar membrane
- stimulated hair cells (receptors) fire off nerv impulses to the brain along the auditory ner

Fact file

FAC

★ The ear becomes less and less sensitive if it is regularly exposed to very loud sounds. At nois discos, you can protect your ears by plugging th with cotton wool.

★ In most humans the ear lobe (pinna) is fixed. Cat and dogs, however, can adjust the pinna and tur towards sources of sound.

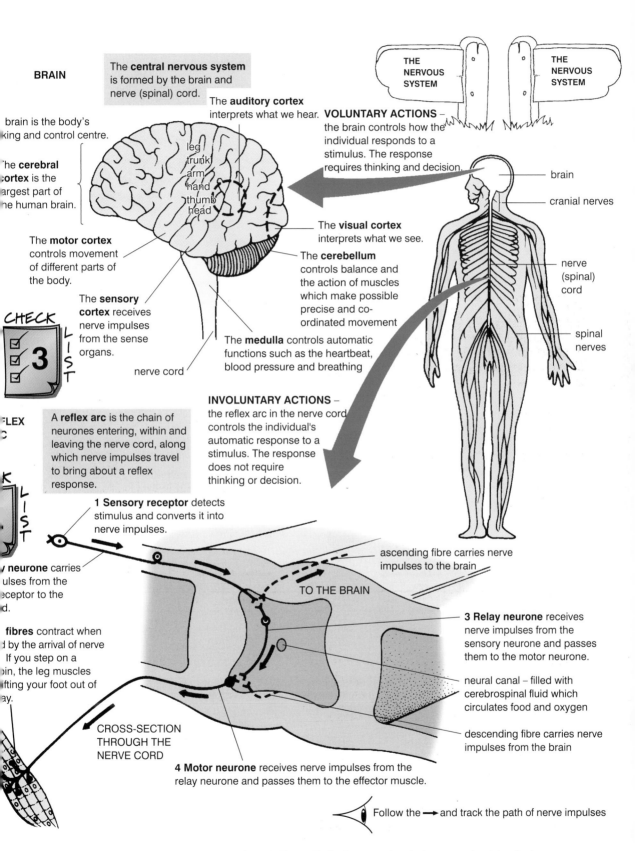

THE NERVOUS SYSTEM

THE NERVOUS SYSTEM

BRAIN

The **central nervous system** is formed by the brain and nerve (spinal) cord.

The **auditory cortex** interprets what we hear.

VOLUNTARY ACTIONS – the brain controls how the individual responds to a stimulus. The response requires thinking and decision.

brain is the body's king and control centre.

leg trunk arm hand thumb head

The **cerebral cortex** is the argest part of he human brain.

The **visual cortex** interprets what we see.

The **motor cortex** controls movement of different parts of the body.

The **cerebellum** controls balance and the action of muscles which make possible precise and co-ordinated movement

The **sensory cortex** receives nerve impulses from the sense organs.

CHECK
☑ ☑ ☑ **3** LIST

nerve cord

The **medulla** controls automatic functions such as the heartbeat, blood pressure and breathing

brain

cranial nerves

nerve (spinal) cord

spinal nerves

INVOLUNTARY ACTIONS – the reflex arc in the nerve cord controls the individual's automatic response to a stimulus. The response does not require thinking or decision.

FLEX

A **reflex arc** is the chain of neurones entering, within and leaving the nerve cord, along which nerve impulses travel to bring about a reflex response.

LIST

1 Sensory receptor detects stimulus and converts it into nerve impulses.

ascending fibre carries nerve impulses to the brain

neurone carries ulses from the eceptor to the d.

TO THE BRAIN

3 Relay neurone receives nerve impulses from the sensory neurone and passes them to the motor neurone.

fibres contract when d by the arrival of nerve If you step on a bin, the leg muscles fting your foot out of ay.

neural canal – filled with cerebrospinal fluid which circulates food and oxygen

descending fibre carries nerve impulses from the brain

CROSS-SECTION THROUGH THE NERVE CORD

4 Motor neurone receives nerve impulses from the relay neurone and passes them to the effector muscle.

Follow the ➡ and track the path of nerve impulses

rones, nerves and the nervous system (checklist on pages 103–104). To simplify the diagram, only a single neurone of each type is shown

The **peripheral nervous system** is formed by the cranial nerves and spinal nerves that join the central nervous system.

NERVES

Neurones are grouped together into bundles called **nerves** which pass to all parts of the body, forming a nervous system.

single neurones

covering around the nerve

CHECK **1** LIST

NEURONE

Neurones are cells specialised to transmit nerve impulses. They build the nervous system.

larger than life

cytoplasm

nucleus

region of the cell body where the nerve impulse starts

cell membrane

dendrites – thin extensions of the cell body that carry nerve impulses to the cell body

axon – long thin extension of the cell body that carries nerve impulses

sheath nucleus

axon ending in muscle

muscle fibre

A sheath formed from a fatty substance called **myelin** wraps around the axon. It boosts the transmission of nerve impulses.

muscl
contra
stimul
the ar
nerve

synapse   neurone   neurone   neurone

synapse

SYNAPSE

Synapses are minute gaps that separate neurones one from another.

Follow the ➝ and track the path of nerve impulses

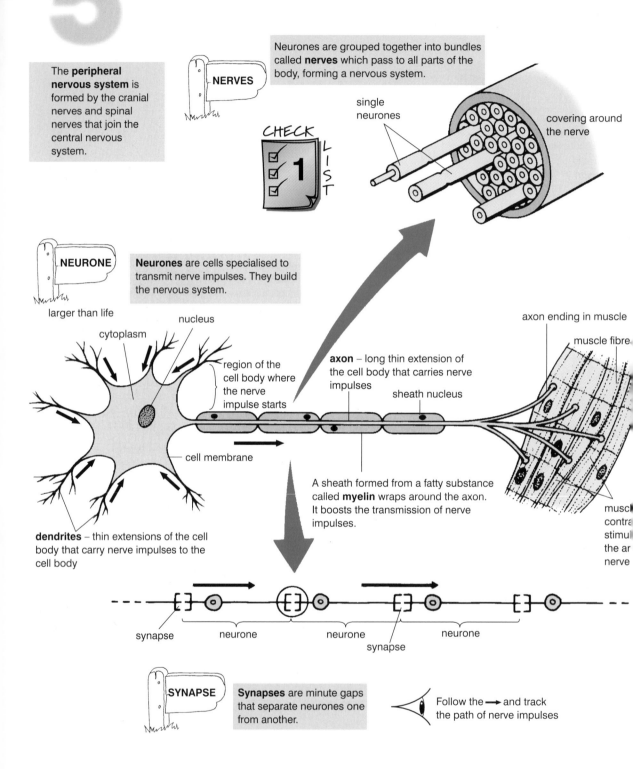

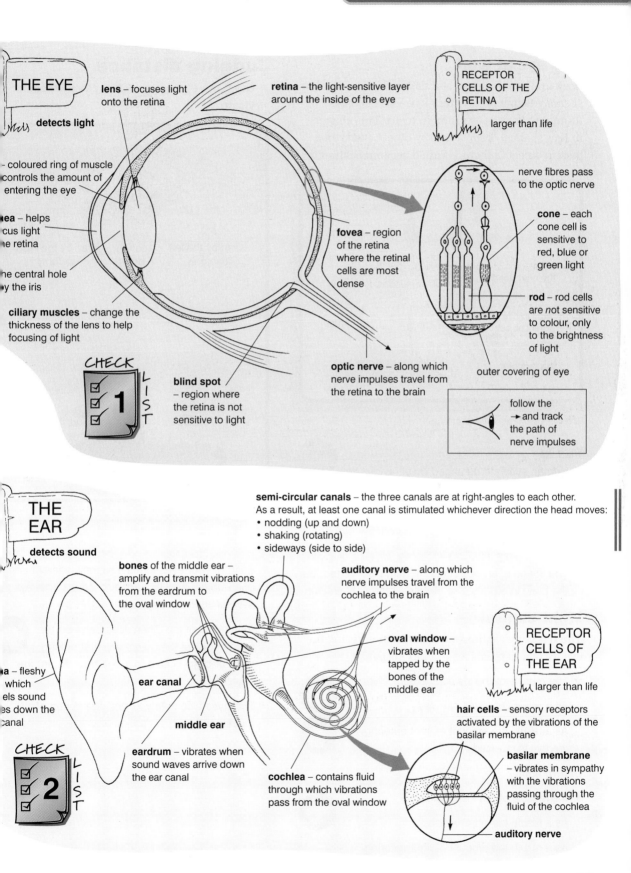

THE EYE

detects light

lens – focuses light onto the retina

retina – the light-sensitive layer around the inside of the eye

– coloured ring of muscle controls the amount of entering the eye

ea – helps cus light he retina

he central hole y the iris

ciliary muscles – change the thickness of the lens to help focusing of light

fovea – region of the retina where the retinal cells are most dense

blind spot – region where the retina is not sensitive to light

optic nerve – along which nerve impulses travel from the retina to the brain

CHECK **1** LIST

RECEPTOR CELLS OF THE RETINA

larger than life

nerve fibres pass to the optic nerve

cone – each cone cell is sensitive to red, blue or green light

rod – rod cells are *not* sensitive to colour, only to the brightness of light

outer covering of eye

follow the → and track the path of nerve impulses

THE EAR

detects sound

semi-circular canals – the three canals are at right-angles to each other. As a result, at least one canal is stimulated whichever direction the head moves:
• nodding (up and down)
• shaking (rotating)
• sideways (side to side)

bones of the middle ear – amplify and transmit vibrations from the eardrum to the oval window

auditory nerve – along which nerve impulses travel from the cochlea to the brain

a – fleshy which els sound es down the canal

ear canal

middle ear

eardrum – vibrates when sound waves arrive down the ear canal

cochlea – contains fluid through which vibrations pass from the oval window

oval window – vibrates when tapped by the bones of the middle ear

RECEPTOR CELLS OF THE EAR

larger than life

hair cells – sensory receptors activated by the vibrations of the basilar membrane

basilar membrane – vibrates in sympathy with the vibrations passing through the fluid of the cochlea

auditory nerve

CHECK **2** LIST

The eye at work

Look up from this page and gaze out of the window at some distant object. Your eye lens becomes thinner to keep your vision in focus. This change in lens shape to keep a nearby object and then a distant object in focus is called **accommodation**.

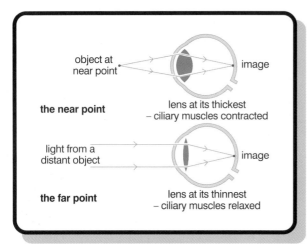

Accommodation keeps objects in focus

Fact file

★ Human eyes are damaged by ultraviolet light. However, insects' eyes can see in ultraviolet light.

★ A normal eye can see clearly any object from far away (at the **far point**) to 25 cm from the eye (the **near point**).

★ The image of an object on the retina is inverted, but the brain interprets it so you see it the right way up.

Light control

The **iris** controls the amount of light entering the eye. Bright light causes a **reflex response**:

★ The muscle of the iris contracts.

As a result, the pupil narrows.

As a result, the amount of light entering the eye is reduced.

In dim light:

★ the muscle of the iris relaxes.

As a result, the pupil widens.

As a result, the amount of light entering the eye is increased.

Judging distance

★ Using two eyes to see with is called **binocular vision**.

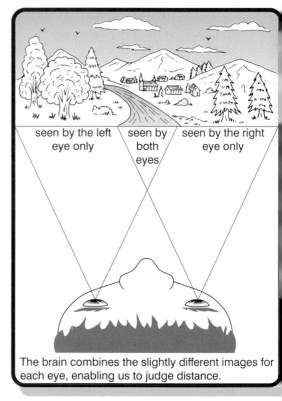

The brain combines the slightly different images for each eye, enabling us to judge distance.

The brain combines the slightly different images for each eye, enabling us to judge distance

★ Each of your eyes forms a slightly different image of an object than its partner because your eyes are several centimetres apart.

As a result of the differences between the two images of the object you are looking at, your brain is able to judge how distant the object is.

★ Two eyes also help you to judge speed of movement.

6 Changing levels of performance

preview

t the end of this section you will:

know that muscle fatigue comes from a lack of oxygen and build up of lactic acid

understand that lactic acid is a product of the chemical reactions of anaerobic respiration

be able to explain why pulse rate and breathing rate increase with exercise

know the meaning of 'recovery time'

understand the link between recovery time and physical fitness.

utual friends!

e muscles of the man and g are working hard. At st, **aerobic respiration** in eir muscle cells gives em a flying start.

AEROBIC
RESPIRATION
Pages 91–92.

$$glucose + oxygen \rightarrow carbon\ dioxide + water$$

$$energy\ released = 16.1\ kJ/g\ glucose$$

e man and dog are both panting. However, in ite of rapid breathing and strenuous pumping the heart, oxygen cannot reach the muscles fast ough to supply their needs.

e muscles then switch from aerobic spiration to **anaerobic respiration** ich does *not* use oxygen.

Lactic acid is produced, which collects in the muscles.

$$glucose \rightarrow lactic\ acid$$

$$energy\ released = 0.83\ kJ/g\ glucose$$

Notice that the energy released per gram of glucose is less than in aerobic respiration. As lactic acid accumulates, the muscles become **fatigued** and stop working. The man and dog will be unable to run any further until the lactic acid has been removed from their muscles.

The removal process uses oxygen which is brought in a rush to the muscles of the man and dog as they recover.

★ The raised level of lactic acid in the blood increases the rate and depth of breathing (panting).

As a result more air is breathed in (supplying oxygen) and breathed out (removing carbon dioxide).

★ Rapid beating of the heart brings more blood with its load of oxygen to the muscles.

Training

During exercise, breathing rate, heart rate and the level of lactic acid in the blood increases less in a fit person than in an unfit person.

Training improves the efficiency of the

★ **lungs**: oxygen is absorbed and passes to the blood *faster*. Carbon dioxide is removed from the blood *faster*.

★ **heart**: blood is pumped around the body *faster*.

Fact file

The beating heart propels blood through the arteries, setting up ripples of pressure. The ripples cause a **pulse** as the muscular walls of the arteries expand and relax in time with the heart beat. The pulse rate therefore measures the heart rate.

Remember that during exercise the heart rate of a fit person is lower than the heart rate of an unfit person.

As a result during exercise the pulse rate of a fit person is lower than the pulse rate of an unfit person.

Recovering

The lactic acid which accumulates in the muscles of the man and dog represents an **oxygen debt**. As they recover, the lactic acid is oxidised and the oxygen debt is repaid.

The **recovery time** is a measure of individual fitness. It is the time taken for the

- breathing rate
- heart rate
- level of lactic acid

in the blood to return to *normal*.

A fit person recovers *more quickly* and therefore has a shorter recovery time than an unfit person.

A fit person pays back the oxygen debt *more quickly* than an unfit person

INVOICE

Attn. OWL
Oxygen debt promptly paid with thanks ✓

Words to remember

You have read some important words in this chapter. Here's a list to remind you what the words in green mean.

Aerobic respiration	uses oxygen to release energy from food
Biceps	a muscle attached to the upper and lower arm bones which, when contracted, raises the low arm
Cones and rods	light-sensitive cells which form the retina of the eye. Cones are sensitive to colour and are packe most densely at the centre of the retina. Rods are not sensitive to colour but respond to the brightness of light. They occur mostly near the edges of the reti
Haemoglobin	a protein which fills red blood cells. Haemoglobin is a red pigment (hence 'red blood cells' which readily combines with oxygen
Mucus	a sticky substance produced by the goblet cells of the mucous membrane
Oxidation	a chemical reaction in which oxygen is added to, or hydrogen removed from, a substance
Oxygen debt	the result of lactic acid accumulating in muscles which are respiring anaerobically
Pacemaker	a group of cells in the right atrium of the heart which contr the beating of the heart
Reflex response	an automatic action not controlled by the brain taken in response to a stimulus
Triceps	a muscle attached to the upper and lower arm bones which, whe contracted, straightens the arm

round-up

How much have you improved?
Work out your improvement on page 137.

1 Complete the following paragraph using the words below. Each word may be used once, more than once or not at all.

thin exhalation fat oxygen inhalation moist carbon dioxide exchange alveoli surface area

The uptake of _____ and removal of _____ occur in the _____ of the lungs. These provide a large _____ for efficient gas _____ . They are _____-walled, have an excellent blood supply, are _____ and kept well supplied with air by breathing. _____ takes air into the lungs; _____ removes air from the lungs. [9]

Well done if you've improved. Don't worry if you haven't. Take a break and try again.

2 The different parts of a motor nerve cell are listed in column **A**. Match each part with its description in column **B**.

A parts of a cell	B descriptions
axon	minute electrical disturbance
dendrite	boosts the transmission of nerve impulses
sheath	transmits nerve impulses from the cell body
nerve impulse	carries nerve impulses to the cell body [4]

3 Explain the differences between the following pairs of terms.

a) blind spot and fovea
b) pupil and iris
c) cornea and retina [6]

4 Briefly summarise the relationship between anaerobic respiration, an oxygen debt and the recovery time. How does physical fitness affect recovery time? [7]

Inheritance

How much do you already know?
Work out your score on page 137.

Test yourself

1 In humans, the gene for brown eyes (**B**) is dominant to the gene for blue eyes (**b**).

   a) Using the symbols **B** and **b**, state the genotypes of the children that could be born from a marriage between a heterozygous father and a blue-eyed mother.   [2]

   b) State whether the children are brown eyed or blue eyed.   [2]

2 List the different sources of variation in living things.   [7]

3 Why does the father determine the sex of the baby and not the mother? How does your answer account for the birth of almost equal numbers of boys and girls?   [7]

6.1 Variation

preview

At the end of this section you will:

● **be able to state what a species is**

● **understand the difference between continuous variation and discontinuous variation**

● **be able to identify the sources of variation**

● **know that variation is either inherited or acquired.**

Names

An organism is given a name consisting of two parts:

★ the **genus** is the first part

★ the **species** is the second part

The genus and species names identify the individual organism. For example, humans belong to the genus *Homo* and have the species name *sapiens*; barn owls are called *Tyto alba*.

That's me folks!

Notice that the genus name begins with a capital letter, the species name begins with a small letter and that the whole name is printed in italics.

What is a species?

Individuals which can sexually reproduce offspring, which themselves are able to reproduce belong to the same species. Hybrids are offspring of members of closely related species. Animal hybrids are often sterile and cannot reproduce.

Fact file

★ A mule is the hybrid offspring from a mating between a horse and a donkey.

Variation within a species

Look closely at your family, friends and classmates. Notice the differently coloured hair and eyes, and the differently shaped faces. We show **variations** in the different characteristics that make up our physical appearance (**phenotype**).

GENOTYPE and PHENOTYPE Page 115.

Variation arises from **genetic** causes.

Sexual reproduction (see pages 71–72) involves the fusion of the nucleus a **sperm** with the nucleus of the **egg**. The fertilisation recombines the genetic material fr each parent in new ways within the **zygote**.

Mutations arise as a result of changes in **chromosomes** or **genes** and are inherited by offspring (see page 117). **Ionising radiation** and some **chemicals** increase the probability of mutation.

Variations that arise from genetic causes are inherited from parents by their offspring, who pass them on to their offspring, and so on from generation to generation. Inherited variation is the raw material on which **natural selection** ac resulting in **evolution**.

Variation also arises from **environmental** cause Here, 'environmental' means all the external influences affecting an organism, for example:

utrients in the food we eat and minerals that
ants absorb in solution through the roots. In
any countries, children are now taller and heavier,
ge for age, than they were 50 years ago because
improved diet and standards of living.

rugs, which may have a serious effect on
ppearance. **Thalidomide** was given to pregnant
omen in the 1960s to prevent them feeling sick
nd help them sleep. The drug can affect the
evelopment of the fetus and some women who
ere prescribed thalidomide gave birth to seriously
eformed children.

emperature affects the rate of enzyme-controlled
hemical reactions. For example, warmth increases
e rate of photosynthesis and improves the rate of
rowth of plants kept under glass.

hysical training uses muscles more than normal,
creasing their size and power. Weightlifters
evelop bulging muscles as they train for their
oort.

ations that arise from environmental causes
not inherited, because the sex cells are not
cted. Instead the characteristics are said to be
uired. The fact that the weightlifter has
eloped bulging muscles does not mean that his
er children will have bulging muscles – unless
take up weightlifting as well!

ntinuous and discontinuous riation

variations shown by some characteristics
spread over a range of measurements. All
rmediate forms of a characteristic are
sible between one extreme and the other.

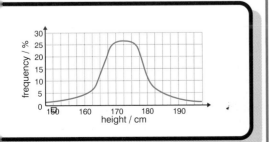

ion in the height of the adult human population – an example of
nuous variation. Variation in weight is another example

We say that the characteristic shows **continuous variation**. The height of a population is an example of continuous variation, as shown.

Other characteristics do not show a continuous trend in variation from one extreme to another. They show categories of the characteristic without any intermediate forms. The ability to roll the tongue is an example – you can either do it or you can't. There are no half-rollers! We say that the characteristic shows **discontinuous variation**.

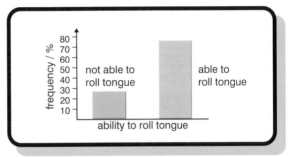

Ability to roll the tongue – an example of discontinuous variation.
Variation in human blood groups (A, B, AB or O) is another example

6.2 What is inheritance?

preview

At the end of this section you will:
● understand genetic terms
● be able to work out the expected outcome of a monohybrid cross
● understand the inheritance of gender
● understand sex-linked inheritance.

Fact file

Gregor Mendel was a monk who lived in the Augustinian monastery at the town of Brünn (now Brno in the Czech Republic). He observed the inheritance of different characteristics in the garden pea, and reported the results of his experiments in 1865. The work established the basis of modern genetics.

The vocabulary of genetics

★ **monohybrid inheritance** – the processes by which a *single* characteristic is passed from parents to offspring, for example flower colour or eye colour

★ **pure breeding** – characteristics that breed true, appearing unchanged generation after generation

★ **parental generation** (symbol **P**) – individuals th are pure breeding for a characteristic

★ **first filial generation** (symbol **F₁**) – the offsprin produced by a parental generation

★ **second filial generation** (symbol **F₂**) – the offspring produced by crossing members of the first filial generation

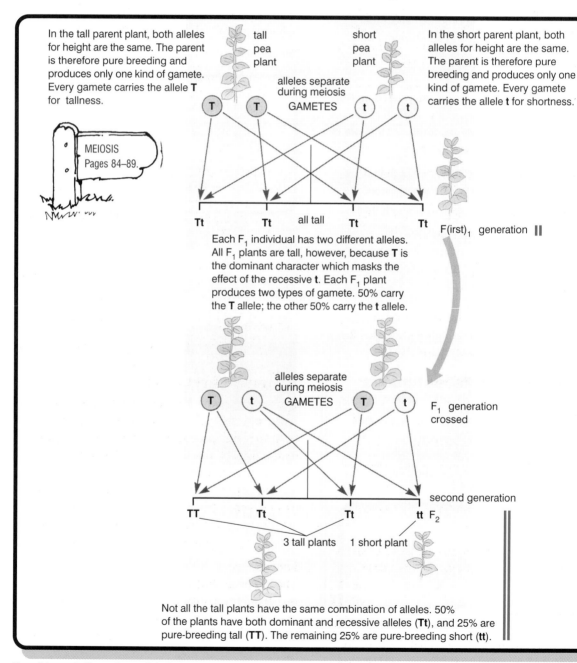

In the tall parent plant, both alleles for height are the same. The parent is therefore pure breeding and produces only one kind of gamete. Every gamete carries the allele **T** for tallness.

MEIOSIS
Pages 84–89.

tall pea plant

short pea plant

In the short parent plant, both alleles for height are the same. The parent is therefore pure breeding and produces only one kind of gamete. Every gamete carries the allele **t** for shortness.

alleles separate during meiosis
T   **T**   GAMETES   **t**   **t**

Tt   **Tt**   all tall   **Tt**   **Tt**   F(irst)₁ generation

Each F₁ individual has two different alleles. All F₁ plants are tall, however, because **T** is the dominant character which masks the effect of the recessive **t**. Each F₁ plant produces two types of gamete. 50% carry the **T** allele; the other 50% carry the **t** allele.

alleles separate during meiosis
T   **t**   GAMETES   **T**   **t**   F₁ generation crossed

second generation
TT   **Tt**   **Tt**   **tt** F₂

3 tall plants   1 short plant

Not all the tall plants have the same combination of alleles. 50% of the plants have both dominant and recessive alleles (**Tt**), and 25% are pure-breeding tall (**TT**). The remaining 25% are pure-breeding short (**tt**).

How alleles controlling a characteristic (height) pass from one generation to the next

ene – a length of DNA which codes for the whole of one protein. Each chromosome carries a number of genes

CHROMOSOMES
Page 84.

llele – one of a pair of genes that control a particular characteristic

omozygote – an individual with identical alleles controlling a particular characteristic. Individuals that are pure breeding for a particular characteristic are **homozygous** for that characteristic

eterozygote – an individual with different alleles controlling a particular characteristic

xpressed – a gene is expressed when a protein is produced from the activity of the gene

ominant – the characteristic which is expressed in an individual heterozygous for the characteristic

ecessive – the characteristic which is expressed only in an individual homozygous for the characteristic

enotype – the genetic make-up (all of the genes) of an individual

henotype – the outward appearance of an individual; the result of those genes of the genotype which are actively expressing characteristics.

me rules of genetics

aired genes controlling a particular characteristic re called alleles.

etters are used to symbolise alleles.

A capital letter is used to symbolise the dominant member of a pair of alleles.

A small letter is used to symbolise the recessive member of a pair of alleles.

The letter used to symbolise the recessive allele is he same letter as that for the dominant allele.

member:

The nucleus of each human body cell contains 46 chromosomes made up of 23 matching pairs of chromosomes. Other organisms (animals and plants) have different numbers of chromosomes but which are always made up of matching pairs.

★ The nucleus of each human sex cell (sperm and egg) contains 23 chromosomes – *half* the number of chromosomes found in the nucleus of each body cell.

★ **Gametes** is a word that refers to both sperm and eggs.

★ Chromosomes carry genes and, of the chromosomes in the fertilised egg (called the zygote), half come from the sperm and half come from the egg.

The process runs:

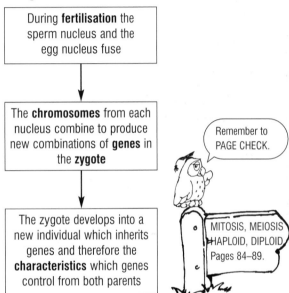

During **fertilisation** the sperm nucleus and the egg nucleus fuse

↓

The **chromosomes** from each nucleus combine to produce new combinations of **genes** in the **zygote**

↓

The zygote develops into a new individual which inherits genes and therefore the **characteristics** which genes control from both parents

Remember to PAGE CHECK.

MITOSIS, MEIOSIS HAPLOID, DIPLOID Pages 84–89.

A monohybrid cross

T is used to symbolise the allele that produces tallness in pea plants, and **t** is used to symbolise the allele that produces shortness. The diagram on the left sets out the results of crosses between tall and short pea plants. Other contrasting characteristics of the pea plant such as seed shape (round or wrinkled), flower colour (purple or white) and pod shape (smooth or wrinkled) are inherited in a similar way.

Fact file
The diagram on the previous page shows that the F_2 generation of a cross between a pure-breeding tall parent plant and a pure-breeding short parent plant predicts a ratio of plants of 3 tall to 1 short. The actual ratio may be different from the predicted ratio because:

- fertilisation is a random process
- some seeds of the F2 generation fail to germinate (grow)
- seedling plants die

Inheritance of sex

The photograph shows the chromosomes that determine the sex of a person. The larger chromosome is the **X** chromosome; the smaller chromosome is the **Y** chromosome. The body cells of a woman carry two X chromosomes; those of a man carry an X chromosome and a Y chromosome.

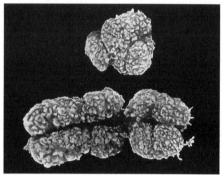

Human sex chromosomes

The diagram below shows how a person's sex is inherited. Notice that

- a baby's sex depends on whether the egg is fertilised by a sperm carrying an X chromosome or one carrying a Y chromosome
- the birth of (almost) equal numbers of girls and boys is governed by the production of equal numbers of X and Y sperms at meiosis.

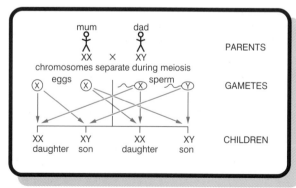

Inheritance of sex in humans

6.3 Genetics and society

preview

At the end of this section you will:

- **know that desirable appearance and increases in production of food (yield) are the result of selective breeding (or artificial selection)**
- **understand that a chromosome mutation causes Down's syndrome**
- **be able to explain that amniocentesis is a technique for detecting chromosome characteristics before birth.**

For centuries we have selected animals and plants for their desirable characteristics and bred from them. The process is called **selective breeding** (or **artificial selection**). For example, dogs have been bred for appearance resulting in a wide variety of breeds.

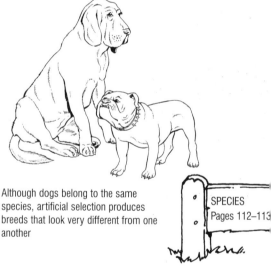

Although dogs belong to the same species, artificial selection produces breeds that look very different from one another

SPECIES
Pages 112–113

In choosing the individuals that are to produce the next generation, humans are choosing the genes (see page 115) controlling the desired characteristics to be passed on to the offspring. The figure shows you the idea.

Apples have other desirable characteristics which have been ignored in the figure (the diagram only tracks the process of selecting apples that are good for eating). For example,

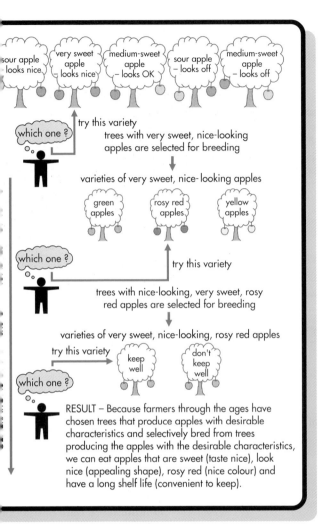

Because of selective breeding we have a range of different types of food selected for sweetness, leanness, appearance, crunchiness and all of the other characteristics of food that we enjoy eating.

Chromosome mutation

Down's syndrome is caused by a chromosome mutation which produces an extra copy of chromosome 21. The photograph shows you the problem.

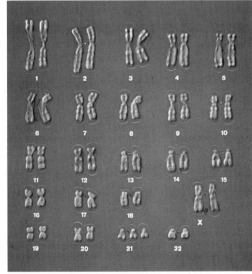

The chromosomes of a person with Down's syndrome. A defect during cell division produces the extra copy of chromosome 21

RESULT – Because farmers through the ages have chosen trees that produce apples with desirable characteristics and selectively bred from trees producing the apples with the desirable characteristics, we can eat apples that are sweet (taste nice), look nice (appealing shape), rosy red (nice colour) and have a long shelf life (convenient to keep).

ive breeding of a variety of apple for eating

ium-sweet apples may be 'good' for cooking, apples 'good' for making jam ... and so on. hoosing different characteristics that make good' eating (or cooking or jamming), rent types of apple have been developed to different tastes.

ners have also selected animals with rable characteristics through selective eding over many generations – for example, stock that are more resistant to disease, e that provide large amounts of milk, pigs produce lean (low fat) meat.

Remember that the nucleus of each human body cell contains 46 chromosomes made up of 23 matching pairs: see page 115.

Sometimes when eggs are produced in a woman's ovaries, the 21st pair of chromosomes fails to separate during meiosis. The eggs, therefore, each have an *extra* chromosome (two chromosome 21s instead of one). If one of the affected eggs is fertilised by a normal sperm then the zygote will carry the extra chromosome (47 in total instead of 46). Development of the zygote results in a baby, and therefore a person, associated with Down's syndrome. The symptoms include speech difficulties and abnormalities of the heart and other organs. The affected person may also have slight mental handicap. Treatment often improves the symptoms.

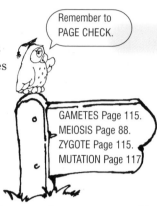

Remember to PAGE CHECK.

GAMETES Page 115.
MEIOSIS Page 88.
ZYGOTE Page 115.
MUTATION Page 117

Some chromosome mutations produce new characteristics which are to our economic advantage. For example, **duplication** of complete sets of chromosomes produces organisms whose cells contain more than the **diploid** number. The organisms are called polyploids.

DIPLOID Page 88.

Many varieties of plants we use for food are polyploids. Wheat, tomatoes and bananas are examples. The multiple sets of chromosomes of modern bread wheat improve yield compared with 'wild' wheat.

Detecting chromosomes

Amniocentesis is a technique that allows doctors to detect genetic disorders in the fetus before birth.

★ A sample of **amniotic fluid** surrounding the fetus developing in the mother's uterus is withdrawn through a thin needle.

★ Living cells shed by the fetus into the amniotic fluid are photographed through a microscope.

★ The chromosomes in the nucleus of each of a sample of fetal cells are checked. If the chromosomes are:

  • *normal*, the parents are reassured that the fetus is healthy

  • *abnormal*, the parents can make a choice about continuing the pregnancy. The decision should take account of moral, ethical and religious questions.

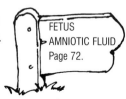

FETUS
AMNIOTIC FLUID
Page 72.

Words to remember

You have read some important words in this chapter. Here's a list to remind you what the words in green mean.

Acquired characteristics	variations in living things th arise from environmental ca
Chromosome mutation	a chromosome error that ari during cell division
Evolution	the process whereby living things are descended from ancestors that have changed through many generations
Hybrid	the offspring of parents whic are not the same species
Natural selection	the process that favours the individuals in a group with t features that best suit them t survive. Less well-suited individuals leave fewer offspring or die before they reproduce
Organism	any living thing
Polyploids	organisms whose cells each contain more than one set of chromosomes
Species	a group of individuals able to mate to reproduce offspring, which themselves are able to mate and reproduce
Sterile	unable to reproduce offsprin
Variation	the differences in characteristics between individuals
Zygote	a fertilised egg

round-up

ow much have you improved?
Vork out your improvement index on page 138.

1 Match each term in column **A** with its correct description in column **B**.

A terms	B descriptions
allele	the processes by which a single characteristic passes from parents to offspring
pure breeding	offspring of the offspring of the parental generation
second filial generation	characteristics that appear unchanged from generation to generation
monohybrid inheritance	one of a pair of genes that control a particular characteristic

[4]

Well done if you've improved. Don't worry if you haven't. Take a break and try again.

2 In a population of 300 goldfish, variations in two characteristics were measured and the results displayed as charts. Chart A shows variation in the length of the fish; chart B shows variation in their colour.

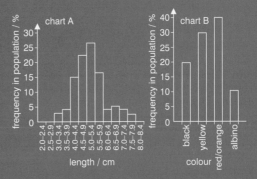

a) Which chart shows
  (i) continuous variation
  (ii) discontinuous variation?
  (iii) Briefly give reasons for your answers. [6]
b) Using chart B, calculate the percentage of yellow goldfish in the population. [1]
c) Albino goldfish are relatively rare. Give a possible genetic explanation for the occurrence of albino goldfish. [1]

3 What is chromosome mutation? Briefly explain how chromosome mutations in different organisms can produce new characteristics which are to human advantage. [4]

Biotechnology

How much do you already know?
Work out your score on page 138.

Test yourself

1 Briefly explain why the production of wine depends on the incomplete breakdown of glucose molecules during anaerobic respiration. [3]

2 Distinguish between the following.
  a) restriction enzyme and ligase (splicing enzyme) [7]
  b) biotechnology and genetic engineering [5]
  c) batch culture and continuous culture [7]

3 Plants are a store of energy. Briefly explain how biotechnology converts the stored energy in plants into fuel. [8]

4 Many types of washing powder contain enzymes. Which types of enzyme are best for washing food-stained clothing? [6]

5 Nitrogen-fixing bacteria, which convert nitrogen in the atmosphere into nitrates, live in swellings on the roots of leguminous plants. Cereal plants do not contain nitrogen-fixing bacteria. How do you think food production and the environment would benefit from biotechnology which manipulates nitrogen-fixing bacteria to live in the roots of cereal plants? [9]

6 a) What is single cell protein? [2]
  b) Single cell protein is produced from bacteria grown on methanol in a fermenter at 40°C.
    (i) Explain the importance of keeping the fermenter at a constant temperature of 40°C. [4]
    (ii) State two forms in which single cell protein is sold. [2]
    (iii) State three advantages of using bacteria to produce protein food. [4]
    (iv) Briefly explain the importance of methanol in the process. [3]

7 Why do you think people might be reluctant to eat food made from microorganisms? [1]

7.1 Living factories

preview

At the end of this section you will:
● **know that we depend on bacteria and yeast to make different foods and drinks**
● **understand the processes that produce wine, bread, cheese and yoghurt**
● **know that fermentation is an anaerobic process**
● **be able to identify the best conditions for the activity of yeast**
● **understand 'souring', 'malting' and 'batch processing'.**

Fact file

★ The word **biotechnology** describes the way we use plant cells, animal cells and microorganisms to produce substances that are useful to us.

For thousands of years humans have exploited microorganisms to make food and drink, using
•   **yeast** to make wine, beer and bread
•   **moulds** to make cheese
•   **bacteria** to make cheese and yoghurt.
The diagram on page 122 traces the processes in the production of wine, bread, cheese and yoghurt.

Notice:

★ *Yeast* cells use **anaerobic respiration** to convert glucose into **ethanol** ('alcohol' in wines and beers) and **carbon dioxide** (the gas that makes bread rise). The reaction is called **fermentation**. Biotechnology exploits a range of fermentation reactions to produce different substances.

For more about the processes of, and comparisons between, aerobic respiration and anaerobic respiration turn to pages 91, 92 and 109.

Moulds: the name given to different types of fungi. Each mould used for making cheese, gives the cheese a particular flavour – for example, 'Stilton' and 'Roquefort' are 'blue' cheeses with characteristic flavours because of the moulds added to them.

Bacteria: bacterial enzymes calayse different reactions which are important steps in the processes of producing cheese and yoghurt. *Lactobacillus* (a type of bacterium – see the diagram on page 122) converts lactose (milk sugar) into lactic acid, which

* separates milk into **curds** and **whey**: an important stage in the production of cheese
* gives yoghurt its sharp flavour.

The lactic acid produced by *Lactobacillus* also gives 'off' milk its **sour** taste. The reactions are examples of **anaerobic fermentation**.

ENZYMES
Pages 88–91.

ct file

Yeasts are fungi that live as single cells. Their name (*Saccharomyces*) means 'sugar fungi'. Yeasts ferment the sugar in rice and barley, making beer, and the sugar in grapes, making wine. For yeast, sugar is food. Fermentation releases energy which keeps yeast alive. We use the ethanol produced to make alcoholic drinks.

Bacteria are also single-celled organisms. However, the cell body is simple compared with the cell body of yeast. There is no distinct nucleus and the structures found in other types of cells are missing.

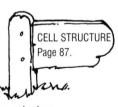
CELL STRUCTURE
Page 87.

ewing the best

ewing beer is a fermentation process (making ne is another version of the process) that pends on yeast.

ewers (people who ke beer) therefore aim provide yeast with the st conditions possible it to make ethanol

MAKING WHISKY
Page 42.

(the 'alcohol' in beer – and lager, wine, whisky and other 'spirits').

★ **Food:** barley seeds contain starch, which is food for the seeds when they begin to grow (**germinate**). Enzymes in barley seeds convert starch to **maltose** (a sugar). Yeast cannot ferment starch. Yeast fements maltose, using it as food and in the process producing ethanol. *Plenty* of barley means *plenty* of food for yeast, which means *plenty* of ethanol for making alcoholic drink.

ENZYMES
Pages 88–91.

Fact file

Germinating the barley seeds in preparation for brewing beer (and making whisky) is called **malting**. The process turns starch into maltose, which can be fermented by yeast.

★ **Temperature:** different types of yeast work best at different temperatures. The optimum temperature for yeasts used to make beer is in the range 12–18 °C; the optimum for lager yeast is 8–12 °C.

OPTIMUM
Page 90.

★ **Sterile conditions:** beer is brewed in large containers called fermenters. Superheated steam at 120 °C is passed through the fermenter and connecting pipelines to kill unwanted microorganisms, which would spoil the beer.

Fact file

Brewing beer is an example of a **batch process**. Once the beer is made, the fermenter is emptied and sterilised before a new batch is made ready for brewing. While the fermenter is out of action no beer is made and no money earned.

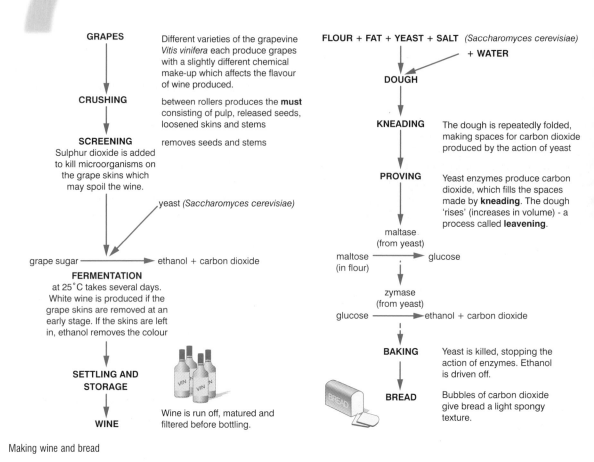

GRAPES

Different varieties of the grapevine *Vitis vinifera* each produce grapes with a slightly different chemical make-up which affects the flavour of wine produced.

↓

CRUSHING

between rollers produces the **must** consisting of pulp, released seeds, loosened skins and stems

↓

SCREENING

removes seeds and stems

Sulphur dioxide is added to kill microorganisms on the grape skins which may spoil the wine.

yeast *(Saccharomyces cerevisiae)*

grape sugar ⟶ ethanol + carbon dioxide

FERMENTATION

at 25°C takes several days. White wine is produced if the grape skins are removed at an early stage. If the skins are left in, ethanol removes the colour

↓

SETTLING AND STORAGE

↓

WINE

Wine is run off, matured and filtered before bottling.

FLOUR + FAT + YEAST + SALT *(Saccharomyces cerevisiae)*

+ **WATER**

↓

DOUGH

↓

KNEADING

The dough is repeatedly folded, making spaces for carbon dioxide produced by the action of yeast

↓

PROVING

Yeast enzymes produce carbon dioxide, which fills the spaces made by **kneading**. The dough 'rises' (increases in volume) - a process called **leavening**.

maltase (from yeast)

maltose (in flour) ⟶ glucose

zymase (from yeast)

glucose ⟶ ethanol + carbon dioxide

↓

BAKING

Yeast is killed, stopping the action of enzymes. Ethanol is driven off.

↓

BREAD

Bubbles of carbon dioxide give bread a light spongy texture.

Making wine and bread

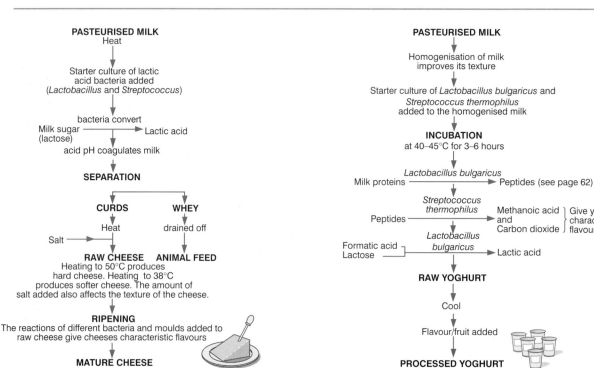

PASTEURISED MILK

Heat

↓

Starter culture of lactic acid bacteria added (*Lactobacillus* and *Streptococcus*)

↓

bacteria convert

Milk sugar (lactose) ⟶ Lactic acid

acid pH coagulates milk

↓

SEPARATION

CURDS   **WHEY**

Heat   drained off

Salt ⟶

RAW CHEESE   **ANIMAL FEED**

Heating to 50°C produces hard cheese. Heating to 38°C produces softer cheese. The amount of salt added also affects the texture of the cheese.

↓

RIPENING

The reactions of different bacteria and moulds added to raw cheese give cheeses characteristic flavours

↓

MATURE CHEESE

PASTEURISED MILK

↓

Homogenisation of milk improves its texture

↓

Starter culture of *Lactobacillus bulgaricus* and *Streptococcus thermophilus* added to the homogenised milk

↓

INCUBATION

at 40–45°C for 3–6 hours

Lactobacillus bulgaricus

Milk proteins ⟶ Peptides (see page 62)

Streptococcus thermophilus

Peptides ⟶ Methanoic acid and Carbon dioxide } Give y⟨ charac⟨ flavour

Lactobacillus bulgaricus

Formatic acid / Lactose ⟶ Lactic acid

↓

RAW YOGHURT

↓

Cool

↓

Flavour/fruit added

↓

PROCESSED YOGHURT

Making cheese and yoghurt

.2 Problems and profit ith waste

preview

At the end of this section you will:
- understand the processes that dispose of sewage
- be able to identify the hazards to health and the environment of untreated sewage.

laking the link

od health depends on the prevention of sease. Hygienic disposal of sewage and waste is particularly important measure which controls e spread of disease.

isposal of sewage

n average a person produces 1.5 litres of urine d faeces each day. Urine and faeces are mponents of sewage; so too are industrial and usehold wastes and the water and grit that ain from roads and paths.

ntreated sewage is a health hazard. It contains icroorganisms causing diseases such as **cholera**, phoid, **poliomyelitis** and **diphtheria**. It also tracts insects which help to spread disease.

her microorganisms help us treat sewage by eaking it down into harmless and even useful bstances. This treatment occurs in sewage orks. The sequence of processes is shown in the agram.

otice in the diagram that ration adds air (and erefore oxygen) to the robic processes that eak down the organic aterial in sewage into mpler substances.

AEROBIC RESPIRATION Pages 91–92.

Trickling filter process: liquid sewage is sprayed on to **filter beds** of clinker, gravel or moulded plastic which provide air spaces. Different types of microorganism cover the filter bed, each type breaking down different components in the sewage.

★ **Activated sludge process:** air is bubbled through tanks containing liquid sewage. The added oxygen promotes the aerobic activity of different types of microorganism, breaking down different components in the sewage.

Fact file

Different types of microorganism are needed for complete treatment of sewage because of the many different types of waste material present. Each type of microorganism breaks down a particular type of waste material.

Damage to the environment

Untreated sewage entering streams, rivers and the sea causes the water to become richer and richer in nutrients. The process is called eutrophication.

★ Water plants increase and grow.

★ Algae increase in number, clouding the water in a greenish scum – an **algal bloom**.

When the plant material and algae die, bacteria decompose the organic matter and multiply.

As a result the bacteria use up the oxygen in the water.

As a result wildlife dies through lack of oxygen.

As a result the water becomes black and unable to support wildlife.

Decomposition

The treatment of sewage is an example of **decomposition** in action. **Bacteria** and **fungi** release enzymes which digest dead organic material. The products of digestion are:

★ absorbed by the bacteria and fungi and are a source of energy for their own growth

★ released into the environment as **nutrients**.

As a result the elements the nutrients contain are available for the growth of new plants. Animals obtain nutrients by eating plants and/or other animals.

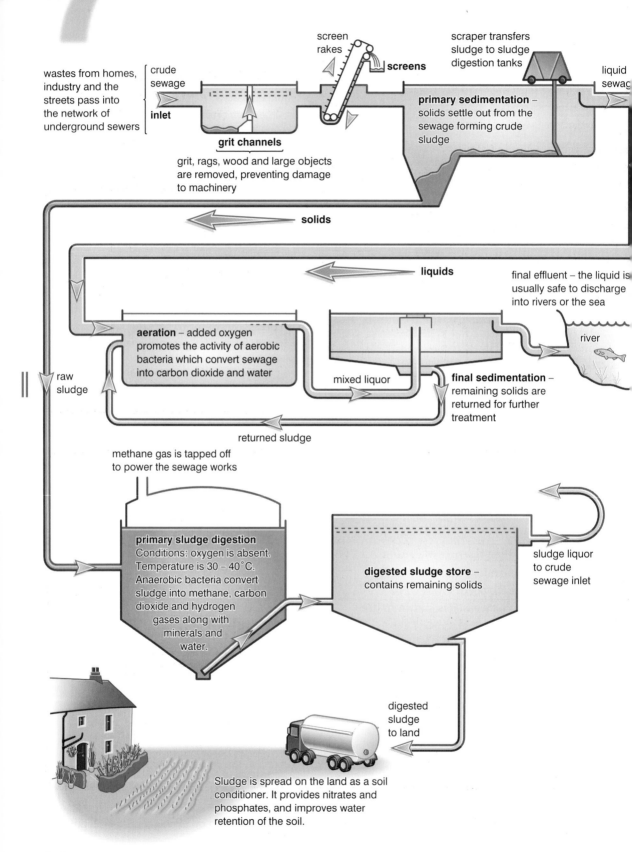

screen rakes

screens

scraper transfers sludge to sludge digestion tanks

liquid sewag

wastes from homes, industry and the streets pass into the network of underground sewers

crude sewage

inlet

primary sedimentation – solids settle out from the sewage forming crude sludge

grit channels

grit, rags, wood and large objects are removed, preventing damage to machinery

solids

liquids

final effluent – the liquid is usually safe to discharge into rivers or the sea

river

aeration – added oxygen promotes the activity of aerobic bacteria which convert sewage into carbon dioxide and water

mixed liquor

final sedimentation – remaining solids are returned for further treatment

raw sludge

returned sludge

methane gas is tapped off to power the sewage works

primary sludge digestion
Conditions: oxygen is absent. Temperature is 30 – 40°C. Anaerobic bacteria convert sludge into methane, carbon dioxide and hydrogen gases along with minerals and water.

digested sludge store – contains remaining solids

sludge liquor to crude sewage inlet

digested sludge to land

Sludge is spread on the land as a soil conditioner. It provides nitrates and phosphates, and improves water retention of the soil.

Treating sewage

ct file

e most common elements which make
 living matter are carbon, hydrogen, nitrogen,
ygen, phosphorus and sulphur.

member the mnemonic **CHNOPS**.
e letters are the chemical
nbols for the elements.

CHNOPS
Page 60.

composition puts back large
ounts of different elements
o the environment. These
ements are the raw materials for making the
rbohydrates, fats and proteins new
nerations of living things use to build bodies.
composition therefore
ycles elements between
e dead and new life. The
ycling of nitrogen and
rbon illustrates the
ocess.

NITROGEN CYCLE
Page 21.

Treatment of water

Four-fifths of the diseases in the Developing
(Third) World are caused by people drinking dirty
water. Providing water which is safe to drink is
the single most effective measure for controlling
the spread of disease. The process runs:

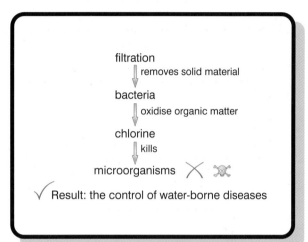

filtration
⇓ removes solid material
bacteria
⇓ oxidise organic matter
chlorine
⇓ kills
microorganisms ✕ ☠

✓ Result: the control of water-borne diseases

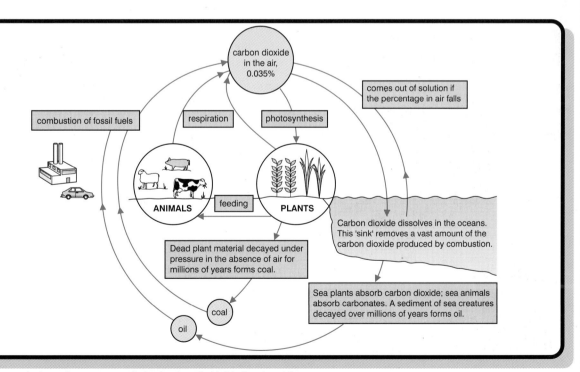

carbon cycle shows those processes that put carbon dioxide into the air and those
remove carbon dioxide from the air. The nitrogen cycle is shown on page 21

7.3 Using waste

preview

At the end of this section you will:

- **understand how waste is converted into useful materials**
- **be able to summarise the advantages of biofuel compared with fossil fuel**
- **know that microorganisms are processed to make food**
- **understand why precautions are taken when handling microorganisms.**

Most household waste in the UK is dumped into large holes in the ground. The operation is called **landfill**. Bacteria naturally present in the soil decompose the waste. When the hole has been filled, it is covered with soil and returned to agricultural use.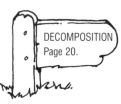

DECOMPOSITION Page 20.

There are problems with landfill.

★ Decomposition of waste produces **methane** gas, which is inflammable. If the gas accumulates, it may explode.

★ Water percolating through the waste dissolves salts of lead, cadmium and copper. These heavy metals are very poisonous and may leak from landfill sites into domestic water supplies.

Other ways of disposing of waste include **incineration** and **recycling**.

Remember that:

★ **methane** gas produced by bacteria fermenting the waste material (dumped in landfill sites) is a problem that can be turned to human advantage.

The gas is piped off for industrial and domestic use as **fuel**

★ **food** is a useful product produced from yeast grown on waste sugar solution. The sugar is food for the yeast which multiplies rapidly. The yeast is harvested and turned into high-protein **animal food** and tea-time favourites like 'Marmite'.

Fuel and food are examples of products produced by the activities of microorganisms **upgrading** waste.

UPGRADING?

Biofuel

Plants trap light energy to produce sugars by **photosynthesis**. Plant material therefore represents a *store of energy* which **fermentation** converts into **fuels** (biofuel) such as ethanol. Countries that lack oil (fossil fuel) resources, such as Brazil, have developed **gasohol** programmes. Sugar cane grows rapidly in the warm sunny environment, and is used as a substrate which yeast ferments to produce ethanol. Most yeasts die when the concentration of ethanol is more than 15%. Sugar cane waste that has not been fermented (called **bagasse**) is burnt to provide the heat to distil off ethanol from the solution, and the gasohol fuel produced is 96% ethanol. After adjustments to carburettors and fuel pumps, car engines can run

PHOTOSYNTHESIS Pages 52–54. FERMENTATION Page 121.

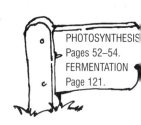

Answer
- waste is *low-grade material with little* energy or nutrient value
- fuel and food are *high-grade materials with high energy* (fuel) and nutrient (food) values.

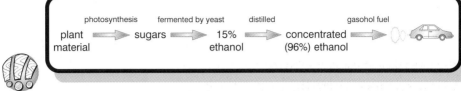

photosynthesis → fermented by yeast → distilled → gasohol fuel

plant material → sugars → 15% ethanol → concentrated (96%) ethanol → car

pure ethanol or an
anol–petrol mixture.
e sequence runs:

member that
thane is also a fuel
oduct of fermentation.

e combustion (burning) of fuel releases energy.
also releases substances (**pollutants**) that
mage the environment. The advantages of
fuel compared with fossil fuel are:

supply: Biofuel is a **renewable** source. Fermentation
processes replace biofuel as it is used. Fossil fuel is
non-renewable. Supplies of fossil fuel are limited
and once used up cannot be replaced

conservation: using biofuel helps supplies of fossil
fuel last longer, giving time for the development of
alternative sources of fuel

lean burn: compared with fossil fuel, burning biofuel
reduces the release of pollutants.

ating microorganisms

w can we produce enough food to feed the
rld's growing population? Eating
croorganisms produced by biotechnology is an
tion.

Microorganisms double their mass within hours.
Plants and animals take weeks.

Microbial mass is at least 40% protein.

Microorganisms have a high vitamin and mineral
content.

gh protein food produced from microorganisms
called **single cell protein** (**SCP**).

oducing SCP

fferent nutrients, such as glucose syrup, waste
m papermaking and fruit pulp, are used to
ow SCP microorganisms in huge fermenters
ich are run continuously for months at a time
ntinuous culture).
trients are replaced as
ey are used up, and
mperature and pH are
refully controlled.

BATCH CULTURE Page 121.

Microorganisms are harvested at regular intervals
and processed into SCP. 'Quorn', made from the
mould *Fusarium graminearum*, is an example.
The food is called **mycoprotein** and, unlike meat,
it is high in fibre and free of cholesterol. Quorn is
made into soups, biscuits and drinks as well as
substituting for different meats.

Inside fermenters making SCP, **warmth**, **food**,
moisture and **oxygen** are controlled to make
conditions ideal for the rapid **asexual
reproduction** of the microorganisms harvested as
food. Think of the sequence of numbers:

2 4 8 16 32 64 128 256 512 1024 … and so on

Notice that the next number in the sequence is
double the previous number. If a population of
microorganisms growing in
a fermenter increases in
this way, it means that each
generation is double the
size of the previous
generation.

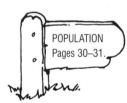

POPULATION Pages 30–31.

Remember that under ideal conditions some
microorganisms can reproduce asexually every
20–30 minutes:

divides → divide → divide
1 cell → 2 cells → 4 cells … and so on

No wonder, with such rapid growth using cheap
nutrients, microorganisms are commercially
attractive for the production of protein-rich foods
– for *animals* as well as humans.

Microorganisms used for industrial processes
and laboratory work need careful handling.

★ Unwanted microorganisms can *contaminate*
  • industrial processes spoiling the product
  • laboratory experiments spoiling the results of
    scientific analysis and research.

★ Microorganisms which cause disease (**pathogens**) may represent a *health hazard* to people working on industrial processes and in laboratories.

Precautions for industry

★ *Superheated steam* at 120 °C is pumped through the fermenter and its pipelines.

As a result spores produced by bacteria and fungi, which may spoil the product and/or be a hazard to health, are destroyed.

Precautions for laboratories

★ *Protective* laboratory coats/face masks prevent bacteria from contaminating clothes and face.

★ *Washing* hands *before* work and *after* work removes microorganisms from the skin.

★ *Swabbing* working surfaces prevents the build up of microorganisms in the laboratory.

★ *Flaming* wire loops and necks of culture bottles makes sure that the microorganisms under investigation are not contaminated with unwanted microorganisms.

★ *Autoclaving* heats laboratory equipment before use so that unwanted microorganisms are killed. It also kills all microorganisms on equipment after the equipment has been used. The precaution makes sure that any unwanted microorganisms that may have contaminated experiments are killed.

Remember to use only cultures of microorganisms that are described as *safe*. **Do not** culture microorganisms from other sources such as air, soil or water. The sources may contain microorganisms that are a hazard to health.

7.4 Reprogramming microorganisms

preview

At the end of this section you will:
- **know that genetic material (genes) can be transferred from one type of organism to another type of organism**
- **understand how insulin is produced by genetically engineered bacteria**
- **be able to compare selective breeding with genetic engineering.**

In the 1970s scientists developed the technique of **genetic engineering**, which introduced the modern era of biotechnology. New methods of manipulating genes became possible because of the discovery of different enzymes in bacteria.

★ **Restriction enzymes** cut DNA into pieces, making it possible to isolate specific genes.

★ **Ligase** (splicing enzyme) allows desirable genes to be inserted into the genetic material of host cells. Yeast cells and the cells of bacteria are often used as host cells for the insertion of desirable genes.

Using genetic engineering we can create organisms with specific genetic characteristics, such that they produce substances that we need and want. The microorganisms are cultured in a solution containing all the substances they require for rapid growth and multiplication inside huge containers called **fermenters**. In this way, medicines, foods and industrial chemicals can be made on an industrial scale. The diagram shows how genetically engineered **insulin** is made.

INSULIN
Page 129.

Remember that the products of biotechnology come from the action of **genes** producing useful substances.

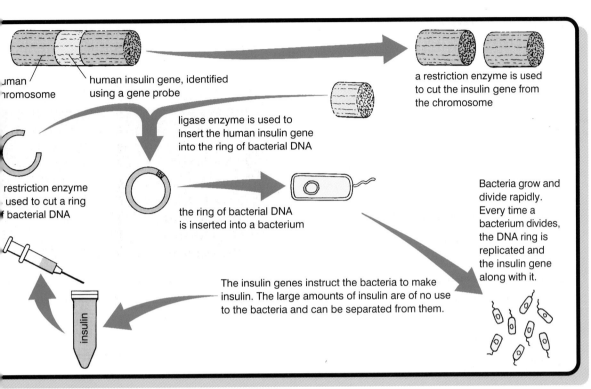

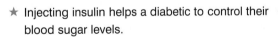
...ng genetically engineered insulin

...example, insulin is produced because of the ...vity of the human insulin gene inserted into ... genetic material of bacteria. The ...omosomal material controls the activities of ... bacteria.

...n greater amounts of a useful substance (e.g. ...lin) can be produced if more than one copy ... desirable gene is inserted into the genetic ...erial of the host cell.

...netimes a cell's own genes can be copied to ...ed up a useful process. For example, some ...es of bacteria have a gene for breaking down ... Inserting more copies of the gene means that ...'oil gene plus' bacteria can clear oil spills ...er than their 'normal' cousins.

...sulin facts

...nsulin is a hormone
...produced by the
...pancreas. It reduces the
...evel of sugar circulating
...n the blood.

PANCREAS
Page 65.

★ A **diabetic** is a person whose pancreas does not produce enough insulin.

As a result the person's blood sugar increases to a level that may endanger health.

★ Injecting insulin helps a diabetic to control their blood sugar levels.

★ The number of diabetics grows because people are living longer (the older the person, the greater the risk of diabetes).

As a result there is an ever-increasing need for insulin.

★ The availablity of insulin used to be limited because we depended on animals (cows and pigs) as the source of supply. Now, genetic engineering produces enough human insulin to meet demand.

7

Selective breeding v Genetic engineering

For centuries we have selected animals and plants for their desirable characteristics and bred from them. The process is called **selective breeding**.

SELECTIVE BREEDING Page 116.

For example, crops have been selectively bred to yield more grain and fruit; livestock to yield more milk and meat. Genetic engineering has opened up new possibilities.

★ Desirable characteristics can be developed in organisms engineered with genes from another type of organism. Selective breeding cannot transfer genes from one type of organism to a different type of organism.

★ New organisms with desirable characteristics can be genetically engineered in one or two generations. Selective breeding takes much longer.

Are genetically engineered organisms harmful?

Answer People are worried about the **unpredictable** effects of genetically engineered organisms which are sources of genetically modified (GM) food. There may be risks to wildlife and human health. More research will help us to make sensible decisions about the way forward.

7.5 Enzymes, detergents and antibiotics

preview

At the end of this section you will:

- know that enzymes are useful industrial catalysts
- understand the advantages of immobilized enzymes
- know why biological detergents help make clothes cleaner
- be able to identify how different antibiotics attack bacteria which cause disease
- understand the problems of bacteria becoming resistant to antibiotics.

Enzymes

Enzymes are useful **industrial catalysts** for the following reasons.

ENZYMES Pages 88–91.

★ Only a particular reaction is catalysed by an enzyme making it easier to collect and purify the products.

★ Enzyme activity is high at moderate temperature and pH.

★ Only small amounts of enzyme are required.

★ The enzyme is not used up in the reaction.

Most industrial enzymes come from microorganisms grown in nutrient solution inside large fermenters. The enzymes secreted into the nutrient solution are filtered off, concentrated and packaged for sale as liquids or powders.

Enzymes have a range of uses.

★ **Industrial** – food production, leather-making, brewing and the manufacture of detergents.

★ **Medical** – diagnosis and treatment.

★ **Analysis** – environmental pollution and crime detection.

mobilised enzymes

ymes may be bonded to different insoluble
erials which support them. They are then
ed **immobilised enzymes**, and their
antages are that they are

easily recovered and can be re-used
active at temperatures that would destroy
unprotected enzymes
not diluted and therefore do not
contaminate the product.

nobilised enzymes are vital components of
erent types of **biosensor**.

nobilised enzymes also make it easier to
duce products by **continuous-flow
cessing**. Substrate flowing over immobilised
yme inside the fermenter is converted into
duct. The product is drawn off from the
nenter as an *ongoing process*. Nutrients are
laced as they are used.

What are the advantages of continuous-flow
processing compared with batch processing
(see pages 121 and 127)?

swer

Removing product as it is made means that
costly separation of product from nutrient and
enzyme is not necessary. Separation of product
s usually a stage of batch processing.

Continuity means that the process can run for a
ong time without the cost of
• stops for cleaning
• loss of product while cleaning.

etergents

tergents remove stains and dirt from clothes.
he stains and dirt contain biological
terial, then **biological detergents** make it
ier to wash out the offending marks.

logical detergents contain enzymes which
est biological material.
member the enzymes which digest the
nponents of food – carbohydrates,

proteins and/or lipids. The same enzymes also
break down the biological material in stains
and dirt. They are produced by bacteria into
which the genes which make the enzymes have
been inserted (engineered).

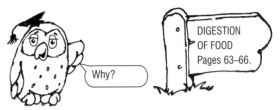

DIGESTION
OF FOOD
Pages 63–66.

Why?

Answer

The biological material in stains and dirt consists
of *carbohydrates*, *proteins* and/or *lipids*, which are
not only compounds of food but also of blood,
grass and other substances that mark clothes.

Digestive enzymes are ingredients of biological
detergents. Their activity makes it easier to
wash out of clothes stains and dirt that contains
biological material.

Biological detergents have advantages over non-
biological detergents. They work best at low
temperatures therefore saving energy (and
money) on heating water. Low temperatures
also mean that the enzymes in detergents are
not destroyed.

Antibiotics

The word **antibiotic** means 'against life'.
Antibiotic drugs attack bacteria which cause
disease:

• **bactericides** such as penicillin kill bacteria
• **bacteristats** such as tetracycline prevent
  bacteria from multiplying.

The diagram at the top of the opposite page shows
how different antibiotics affect bacteria.

Fact file

Three scientists were involved in the
discovery and development of penicillin.

★ In 1928 **Alexander Fleming** noticed that the mould
Penicillium notatum killed bacteria. He isolated the
active substance and called it **penicillin**.

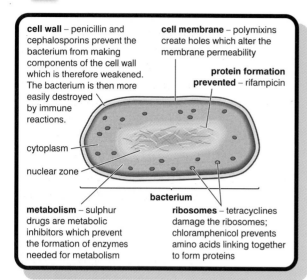

cell wall – penicillin and cephalosporins prevent the bacterium from making components of the cell wall which is therefore weakened. The bacterium is then more easily destroyed by immune reactions.

cytoplasm

nuclear zone

cell membrane – polymixins create holes which alter the membrane permeability

protein formation prevented – rifampicin

bacterium

metabolism – sulphur drugs are metabolic inhibitors which prevent the formation of enzymes needed for metabolism

ribosomes – tetracyclines damage the ribosomes; chloramphenicol prevents amino acids linking together to form proteins

Antibiotic drugs at work. Ribosomes are the places within the cell where proteins are made.

★ In 1938 **Howard Florey** and **Ernst Chain** began to develop methods to produce enough penicillin for clinical trials.

★ By 1944 penicillin was in large-scale production, following the move of production from Britain to the USA at the onset of the Second World War.

Resistance

Bacteria can become **resistant** to a particular antibiotic, and that antibiotic becomes less effective for the treatment of disease.

★ The dosage of a drug to which bacteria are becoming resistant has to be gradually increased as symptoms continue.

As a result, the drug becomes increasingly inefficient as resistance develops.

As a result, the drug may become poisonous to the patient.

How can we slow the spread of resistance?

Answer

★ Avoid using antibiotics by practising good hygiene to prevent the spread of infection.

★ Use antibiotics sparingly when drugs are neede to treat infection.

★ Finish a prescribed course of antibiotics.

★ Reduce the antibiotics given to farm animals.

How do doctors deal with resistant infections?

Answer

★ Different types of antibiotic are prescribed to treat different diseases, and drugs are switched if resistance develops.

★ Research scientists work continually to develop new antibiotics.

However, the race to develop new drugs as bacteria develop resistance to existing ones is close. The race is an example of the 'Red Quee Effect'. Find out what the Red Queen says to Alice in *Through the Looking Glass* by Lewis Carroll.

Words to remember

You have read some important words in this chapter. Here's a list to remind you what the words in green mean.

Biofuel — fuel obtained by using yeast t ferment the sugars in plant material into ethanol

Culture — the production of large numb of organisms (usually microorganisms)

Detergent — a substance (not soap) used w water to remove dirt

Eutrophication — the process which leads to excessive plant (algal) growth water enriched with nutrient (e.g. nitrate). When the excess plant (algal) material dies, the activities of decomposers reduces the concentration of

dissolved oxygen and fish and other wildlife die from lack of oxygen	
mentation the anaerobic (i.e. without oxygen) reactions of yeast and other microorganisms producing useful substances (e.g. ethanol)	
mones chemicals produced and released by different tissues into the bloodstream	

Spores	single-celled (usually) structures produced by the parent organism as a result of sexual or asexual reproduction. Spores may scatter (disperse) and each is able to germinate and develop into a new individual in favourable conditions (e.g. warm, plenty of moisture)
Substrate	a substance, the reaction of which is catalysed by an enzyme

round-up

How much have you improved? Work out your improvement index on pages 138–139.

1 Distinguish between the processes of kneading, proving and baking in the making of bread. [6]

2 Below is a list of the processes which start with identifying the human insulin gene and result in the production of genetically engineered human insulin. Write the letters in the correct order.

A The human insulin gene is identified.

B Bacteria genetically engineered with the human insulin gene grow and divide rapidly.

C Restriction enzyme cuts open a ring of bacterial DNA.

D Large amounts of insulin are separated from the nutrient solution in which the genetically engineered bacteria are growing.

E The ring of genetically engineered bacterial DNA is inserted into a bacterium.

F Restriction enzyme cuts the human insulin gene from the chromosome.

G Ligase is used to insert the human insulin gene into a ring of bacterial DNA.

H Human insulin is purified and packed. [6]

3 What are the advantages to diabetics of using insulin produced by bacteria into which the human insulin gene has been inserted? [4]

4 How is biotechnology helping to feed the world's growing human population? [5]

5 What is an immobilised enzyme? [3]

6 People might be reluctant to eat SCP made from microorganisms. How do you think SCP might be made more acceptable as food? [4]

7 What contribution do you think that SCP production could make to helping feed the world's growing population? [7]

8 List in correct order the stages in the treatment of sewage in a modern sewage works. [7]

Well done if you've improved. Don't worry if you haven't. Take a break and try again.

… THAT'S IT FOLKS!

Answers

Test yourself (page 16)

Introducing biology

1 a) It would increase (✓).   **b)** It would decrease (✓).

2 a)

	b)	c)
Movement	✓	✗
Respiration	✓	✗
Sensitivity	✓	✗
Growth	✓	✗
Reproduction	✓	✗
Excretion	✓	✗
Nutrition	✓	✗	[✓ × 10]

d) No – plants do not move from place to place (✓).

3 a) To identify different living things by name (✓).
  b) These characteristics vary too much (✓) even between members of the same group of organisms (✓) to be reliable indicators for identification (✓).

Your score: ☐ out of 17

Round-up (page 18)

Introducing biology

1 a) It would boil away (✓).
  b) It would freeze and form ice (✓).

2 Oxygen (✓)

3 a) Respiration releases energy from food (✓). Gaseous exchange takes in oxygen needed for respiration (✓) and removes carbon dioxide produced by respiration (✓).
  b) Excretion removes the waste substances produced by metabolism (✓). Defecation removes the undigested remains of food (✓).

4

A characteristics		B descriptions
Movement	(✓)	Changing position
Respiration	(✓)	Releasing energy from food
Sensitivity	(✓)	Responding to stimuli
Growth	(✓)	Increasing in size
Reproduction	(✓)	Producing new individuals
Excretion	(✓)	Removing waste substances produced by cells
Nutrition	(✓)	Making or obtaining food

5 The unfamiliar specimen is compared with the descriptions in the key (✓).
The descriptions are followed through (✓) until the description that matches the specimen is found (✓).
The matching description identifies the specimen (✓).

Your score: ☐ out of 19

Your improvement index: $\dfrac{\boxed{}/19}{\boxed{}/17} \times 100\% = \boxed{}\%$

1 Test yourself (page

The biosphere

1

A terms		B descriptions
Biosphere	(✓)	All the ecosystems of the world
Community	(✓)	All the organisms that live in a particular ecosystem
Habitat	(✓)	The place where a group of organisms lives
Population	(✓)	A group of individuals of the same species

2 a) Most animals eat more than one type of plant or other animal (✓). A food web shows the range of different fo eaten (✓).
  b) Plants produce food by photosynthesis (✓). Animals consume this food directly when they eat plants (✓) or indirectly when they eat other animals (✓) which depen plant food (✓).

3 The pyramid of biomass takes into account differences in si (✓) of producers and consumers (✓).

4 a) Correct axes (numbers: y axis; years x axis (2✓); correc plotted (6✓ = 12 × ½ mark for each correct co-ordinate)
  b) 1840–1880 (2✓) (allow ±5 years).
  c) 1900 (✓); 1930 (✓).
  d) Population numbers were stable (✓). (Allow: population numbers fluctuated around a mean (average) number.)

5 To begin with the numbers of microorganisms rapidly increa (✓). The microorganisms use up dissolved oxygen (✓). Mo types of organism die through lack of oxygen (✓). A few ty of organism survive (✓). Their numbers increase (✓).

Your score: ☐ out of 30

1 Round-up (page

The biosphere

1 Physical *or* abiotic (✓), environment (✓), living *or* biotic (✓ community (✓), habitats (✓).

2 a) The non-living part of an ecosystem (✓).
  b) The amount of light affects the rate of photosynthesis (✓ and therefore the amount of plant growth under the canc layer (✓). This in turn affects the animals that depend o plants for food and shelter (✓) and so on along the food chain (✓).

3 a) 4 (✓)
  b) Water weed (✓)
  c) Water weed makes food by photosynthesis (✓).
  d) Tadpoles (✓)
  e) Tadpoles eat water weed (accept plants) (✓).
  f) Minnows (✓) and perch (✓).
  g) Minnow and perch eat tadpoles (accept meat) (✓).

) Saw wrack (✓).

) Dog whelks (✓).

) The biomass of dog whelks would decrease (✓). The biomass of saw wrack would increase (✓).

) 1% of 400 000 KJm^{-2} of radiant energy is 4000 KJm^{-2} (✓). 10% of 4000 KJm^{-2} of stored energy passes on to the rabbit (✓). Answer: 400 KJm^{-2} (✓).

) Feeding transfers energy from one link in the food chain to the next (✓). The transfer is never 100% efficient (✓). Energy is lost at each link through life's activities (✓). The amount of food energy available to the next link in the food chain is therefore less than the previous link (✓). Eventually food energy dwindles to nothing(✓).

our score: ☐ out of 30

our improvement index: $\dfrac{\boxed{}/30}{\boxed{}/30} \times 100\% = \boxed{}\%$

Test yourself (page 40)

e world of plants

) Carbon dioxide (✓) and water (✓).
) Oxygen (✓).

ncrease (✓), osmotic (✓), osmosis (✓), xylem (✓), ylem (✓), evaporation (✓), stomata (✓).

	B
Nectary	Produces a sugar solution
Stigma	Structure to which pollen grains attach
Fruit	Develops from the ovary after fertilisation
Ovule	Contains the egg nucleus
Seed	A fertilised ovule

) Insects are attracted to flowers (✓) and pick up a load of pollen (✓).
) Pollination brings pollen grains from the anthers (✓) to the stigma of a carpel (✓). Fertilisation occurs when one of the male sex nuclei from a pollen grain (✓) fuses with the female egg nucleus in the ovule (✓).

Root (✓), leaf (✓) and stem (✓).

Plants are healthy (✓), the same (✓) and inherit all the parent's desirable characteristics (✓).

Steeping (✓), germination (✓), drying (✓), grinding (✓), production (✓).

Your score: ☐ out of 34

2 Round-up (page 59)

The world of plants

1 Palisade cells (✓), spongy mesophyll cells (✓), guard cells (✓).

2 The cells of the upper surface of the leaf do not contain chloroplasts (✓). Most light therefore reaches the palisade cells (✓) which are packed with chloroplasts (✓). Photosynthesis occurs at a maximum rate (✓).

3 Temperature (✓), light intensity (✓), supplies of carbon dioxide (✓) and water (✓).

4 photosynthesis (✓), translocation (✓), transpiration (✓).

5 Warm (✓), windy (✓), low humidity (✓) and bright sunlight (✓).

6 The stomata close (✓). If the plant continues to lose more water than it gains then its cells lose turgor (✓) and it wilts (✓).

7
Xylem		Phloem
Dead tissue (cells)	(✓)	Living tissue (cells)
Tissue (cells) waterproofed with lignin	(✓)	Tissue (cells) not waterproofed with lignin
Transports water and minerals	(✓)	Transports sugar and other substances
Transport of materials is one way	(✓)	Transport of materials is two ways
Xylem tissue does not have companion cells	(✓)	Phloem tissue has companion cells

Your score: ☐ out of 26

Your improvement index: $\dfrac{\boxed{}/26}{\boxed{}/34} \times 100\% = \boxed{}\%$

3 Test yourself (page 60)

Animal survival

1 a) Carbohydrates (✓), fats (✓), proteins (✓).
  b) Protein (✓)
  c) Fat (✓)
  d) Minerals (✓) and vitamins (✓).

2
A terms		B descriptions
Ingestion	(✓)	Food is taken into the mouth
Digestion	(✓)	Food is broken down
Absorption	(✓)	Digested food passes into the body
Egestion	(✓)	The removal of undigested food through the anus

3 Glomerulus (✓), Bowman's capsule (✓), tubule (✓), collecting duct (✓), ureter (✓), bladder (✓), urethra (✓).

4 The large pointed canines (✓) allow the dog to grip the food firmly (✓). The carnassial teeth (✓) in the upper and lower jaws cut the food (✓).

Your score: ☐ out of 22

3 Round-up (page 80)

Animal survival

1

A nutrients		B test results
Starch	(✓)	Produces a blue/black colour when mixed with a few drops of iodine solution
Glucose	(✓)	Produces an orange colour when heated with Benedict's solution
Fat	(✓)	Forms a milky emulsion when mixed with warm dilute ethanol
Protein	(✓)	Produces a violet/purple colour when mixed with dilute sodium hydroxide and a few drops of copper sulphate solution

2

A enzymes		B roles
Amylase	(✓)	Digests starch to maltose
Pepsin	(✓)	Digests protein to polypeptides
Lipase	(✓)	Digests fat to fatty acids and glycerol
Maltase	(✓)	Digests maltose to glucose

3 Antidiuretic hormone promotes reabsorbtion of water into the body (✓) by making the collecting duct of the nephron more permeable to water (✓).

4 Calcium (✓), keratin (✓), hardest (✓).

5 A = sperm duct (✓), B = urethra (✓), C = scrotal sac (accept scrotum) (✓), D = testis (✓), E = penis (✓).

6 A woodlouse quickly loses water from its body in dry air (✓). Woodlice are often found in leaf litter because it provides a dark and damp environment (✓). Air is saturated with water vapour (humid) (✓). Loss of water from the body is therefore reduced (✓).

Your score: ☐ out of 22

Your improvement index: $\dfrac{\boxed{}/22}{\boxed{}/22} \times 100\% = \boxed{}\%$

4 Test yourself (page 81)

Investigating cells

1

A structures		B functions
Cell membrane	(✓)	Partially permeable to substances in solution
Chloroplast	(✓)	Where light energy is captured
Cell wall	(✓)	Fully permeable to substances in solution
Nucleus	(✓)	Contains the chromosomes

2 a) A plasmolysed cell is one from which water has passed out of the vacuole, out of the cytoplasm, out of the cell membrane and cell wall into the solution outside the cell (accept water has passed out of the cell) (✓). As a result the cytoplasm pulls away from the cell wall (accept cell content disrupted) (✓) and the cell becomes limp (✓). A turgid cell is one which contains as much water as it can hold (✓).

b) A fully permeable membrane allows most substances to pass through it (✓). A partially permeable membrane all some substances to pass through it (✓) and stops othe substances (✓).

3 A group of genetically identical cells (or organisms) (✓).

4 a) Cells die (✓). New cells (✓) which are replicas of the ol cells are produced by mitosis (✓).

b) The daughter cells have the same number of chromoson as the parent cell (✓). The chromosomes in the daughte cells are identical to those in the parent cells (✓).

5 Cells (✓), cells (✓), tissues (✓), an organ (✓), organs (✓), organ(✓).

6 Cellulose in plant cell walls (✓); chitin in insect exoskeletons (✓).

7 ×40 (✓)

8 carbon dioxide (✓).

Your score: ☐ out of 27

4 Round-up (page 9

Investigating cells

1 a) Nucleus (✓), cell membrane (✓), mitochondria (✓), cytoplasm (✓).

b) Cell wall (✓), large vacuole (✓), chloroplasts (✓).

2 Water is taken into the cells (✓) by osmosis (✓). The cells become turgid (✓). The wilted plant (✓) will become upright again (✓) as its cells become turgid following watering.

3 Down (✓), faster (✓), partially (✓), osmosis (✓).

4 Damaged tissues can be replaced by new cells that are ident to the parent cells (✓).

5 C (✓), B (✓), E (✓), D (✓), F (✓), A (✓)

6 For example:

substrate	product
starch	sugars
protein	peptides
hydrogen peroxide	water and oxygen
sugar	starch

Allow other suitable examples (✓ ×3).

Your score: ☐ out of 26

Your improvement index: $\dfrac{\boxed{}/26}{\boxed{}/27} \times 100\% = \boxed{}\%$

Test yourself (page 94)

body in action

a) There are two bronchi, one branching to each lung (✓). Each bronchus branches many times into small tubes called bronchioles (✓).

b) A person has two lungs (✓). Within each lung, bronchioles subdivide into even smaller tubes which end in clusters of small sacs called alveoli (✓).

c) During aerobic respiration, cells use oxygen to oxidise digested food substances (accept glucose) (✓), releasing energy (✓). During anaerobic respiration, cells break down digested food substances (accept glucose) without oxygen (✓). Less energy is released during anaerobic respiration than during aerobic respiration (✓).

d) Breathing takes in (inhales) (✓) and expels (exhales) (✓) air. Gaseous exchange occurs across the surfaces of the alveoli (✓).

The right ventricle pumps blood into the pulmonary artery (✓) on its way to the lungs (✓); the left ventricle pumps blood into the aorta (✓) which takes it around the rest of the body (✓).

A components	B descriptions
Plasma (✓)	Contains dissolved food substances
Red blood cells (✓)	Contain haemoglobin
White blood cells (✓)	Produce antibodies

Receptor (✓), sensory neurone (✓), relay neurone (✓), motor neurone (✓), effector.

a) The eardrum vibrates (✓) in response to sound waves (✓).

b) The bones pass vibrations through the middle ear (✓) and also amplify them (✓).

c) The pinna funnels sound waves down the ear canal (✓) to the eardrum (✓).

d) The hair cells are stimulated by the vibrations of the basilar membrane (✓). They fire off nerve impulses to the brain along the auditory nerve (✓).

Your score: [ ] out of 30

Round-up (page 111)

he body in action

Oxygen (✓), carbon dioxide (✓), alveoli (✓), surface area (✓), exchange (✓), thin (✓), moist (✓), inhalation (✓), exhalation (✓).

A parts of cell		B descriptions
Axon	(✓)	Transmits nerve impulses from the cell body
Dendrite	(✓)	Carries nerve impulses to the cell body
Sheath	(✓)	Boosts the transmission of nerve impulses
Nerve impulse	(✓)	Minute electrical disturbances

3 a) The blind spot is the region of the retina insensitive to light (✓). The fovea is the most sensitive region of the retina, where cone cells are most dense (✓).

b) The pupil is the central hole formed by the iris (✓). The iris is the coloured ring of muscle that controls the amount of light entering the eye (✓).

c) The cornea bends light (✓) and helps to focus light onto the retina (✓).

4 Active muscles respire anaerobically when oxygen cannot reach the muscles fast enough to supply their energy needs (✓). Lactic acid is produced, accumulates and stops the muscles from working (✓). Accumulation of lactic acid in the muscles represents an oxygen debt (✓). During the recovery time, the lactic acid is oxidised (removed from the muscles) and the oxygen debt repaid (✓). Training improves the efficiency of the lungs and heart (✓). Lactic acid is removed more quickly (✓) and the recovery time is therefore shorter (✓).

Your score: [ ] out of 26

Your improvement index: $\dfrac{[\ ]/26}{[\ ]/30} \times 100 = [\ ]\%$

6 Test yourself (page 112)

Inheritance

1 a) **Bb** (✓) or **bb** (✓).

b) 50% of the children would be brown eyed (✓); 50% blue eyed (✓).

2 Genetic recombination (✓) as a result of sexual reproduction (✓), mutation (✓), nutrients (✓), drugs (✓), temperature (✓), physical training (✓).

3 The mother's eggs each carry an X sex chromosome (✓). Of the father's sperm, 50% each carry an X sex chromosome (✓) and 50% each carry a Y sex chromosome (✓). The chance of an egg being fertilised by an X sperm or Y sperm is approximately 50:50 (✓). If the fertilised egg is XX it develops into a girl (✓); if XY it develops into a boy (✓). The birth of almost equal numbers of boys and girls is therefore governed by the production of equal numbers of X and Y sperm (✓).

Your score: [ ] out of 18

6 Round-up (page 119)

Inheritance

1

A terms		B descriptions
Allele	(✓)	One of a pair of genes that control a particular characteristic
Pure breeding	(✓)	Characteristics that appear unchanged from generation to generation
Second filial generation	(✓)	Offspring of the offspring of the parental generation
Monohybrid inheritance	(✓)	The processes by which a single characteristic passes from parents to offspring

2 a) (i) A (✓)
   (ii) B (✓)
   (iii) Chart A shows intermediate lengths of fish (✓) over a range of measurements (✓). Chart B shows categories of colour (✓) without any intermediate forms (✓).
   b) 90% (✓)
   c) Albino fish occur as a result of a mutation of the alleles controlling colour (✓).

3 A chromosome error that arises during cell division (✓). Some chromosome mutations in organisms may duplicate complete sets of chromosomes (✓), producing polyploids (✓). Some polyploid plants produce more food than non-polyploids (✓).

Your score: ☐ out of 16

Your improvement index: $\dfrac{\boxed{}/16}{\boxed{}/18} \times 100\% = \boxed{}$ %

7 Test yourself (page 120)

Biotechnology

1 In the absence of oxygen (✓), yeast breaks down glucose anaerobically (✓) to form ethanol and carbon dioxide (✓).

2 a) Restriction enzyme cleaves (cuts up) lengths of DNA into different fragments (✓) depending on the restriction enzyme used (✓). A particular DNA fragment corresponds to a desired gene (✓). (Accept sensible alternative explanations.) Ligase splices the desired gene from among the fragments of DNA produced by restriction enzyme (✓) into a plasmid vector (✓) which is a loop of bacterial DNA (✓) into which the desired gene is inserted (✓).
   b) Biotechnology uses microorganisms (✓) on a large scale for the production of useful substances (✓). Genetic engineering manipulates genes (✓) to create organisms (✓) with specific genetic characteristics for producing a range of useful substances (✓).
   c) Batch culture produces substances in a fermenter (✓). The fermenter is then emptied of product and nutrient solution (✓) and sterilised (✓) in preparation for the next batch (✓). Continuous culture produces substances over an extended period (✓). Product is drawn off and nutrients replaced as they are used (✓) during an ongoing process (✓).

3 Photosynthesis produces sugar which is a store of energy (✓). Yeasts are used to ferment the sugar (✓), producing ethanol (✓). When the concentration of ethanol exceeds 15% the yeast die (✓). To concentrate the ethanol so that it is a useful fuel (✓), excess water is driven off (accept distilled off (✓) by burning sugar cane waste (accept bagasse) (✓) as a source heat for the distillation process (✓).

4 Carbohydrases (✓) which digest carbohydrates (✓); protease (✓) which digest protein (✓); lipases (✓) which digest fats (✓

5 Nitrogen-fixing bacteria living in the roots of cereals would provide the plants with nitrate fertiliser (✓), increasing production (✓). The need for synthetic nitrate fertilisers would diminish, saving fuel in their production (✓), reducing the amount applied to crops (✓) and therefore reducing the amou of surplus fertiliser which runs off the land into rivers and lake (✓). Eutrophication (accept excessive growth of algae) would be avoided (✓) and wildlife would therefore benefit (accept BO of water would not increase) (✓). Also nitrates would not poll drinking water supplies (✓), reducing the associated hazards health (✓).

6 a) High protein food (✓) produced from microorganisms (✓).
   b) (i) 40 °C is the best temperature for fermentation reaction to take place (✓). Fermentation reactions quickly raise the temperature inside the fermenter (✓). Temperature of more than 60 °C (accept high temperatures) would kill the cell culture (✓). A cooling system is necessary to maintain a temperature of around 40 °C (✓).
   (ii) Marmite (✓) and Quorn (✓) are two examples. (Acce sensible alternative suggestions.)
   (iii) Microorganisms double their mass within hours (acce very quickly) (✓) compared with weeks for plants and animals (✓).
   Microbial mass is at least 40% protein (✓).
   Microbial mass has a high vitamin and mineral content (✓).
   (iv) Methanol is a substrate (✓) which bacteria ferment (accept use as food) (✓) to produce single cell protein (✓).

7 People associate microorganisms with dirt and disease (✓).

Your score: ☐ out of 61

7 Round-up (page 133

Biotechnology

1 Kneading – repeated folding of the dough (✓) makes spaces fo carbon dioxide produced by the action of yeast enzymes on the sugar in the dough (✓).
   Proving – carbon dioxide fills the spaces produced by kneading (✓).
   Baking – kills yeast, stopping the action of enzymes (✓).
   Ethanol produced by yeast fermenting sugars (✓) is driven off (✓).

2 A, F, C, G, E, B, D, H (✓✓✓✓✓✓).

Genetically engineered insulin is cheaper (✓), available in large quantities (✓) and chemically the same as human insulin (✓), preventing a possible immune response to injection of the hormone (✓).

Developments include
- genetically engineering crops to grow in places where at present there is little chance of success (✓)
- altering nitrogen-fixing bacteria so that they can live in the roots of cereal crops (✓)
- designing insecticides produced by bacteria and which are selective for particular insect pests (✓)
- producing plants resistant to disease (✓)
- developing livestock to produce more and better quality meat and milk (✓).

An immobolised enzyme is made by attaching the enzyme to an insoluble support (✓). This means that enzyme is not lost (✓) when the products are collected (✓).

Addition of colours (✓) and flavours (✓); blending with meat (✓) produces acceptable food (✓).

High protein food (✓) which is also high in vitamins and minerals (✓) but low in cholesterol (accept cholesterol-free) (✓) is produced by industrial processes (✓) which occupy a small area of land compared with a farm (✓). Production processes can be closely controlled (✓) compared with crops and livestock which are exposed to a variety of uncontrollable environmental factors (✓).

Grit removal (✓), primary sedimentation (✓), aeration (✓), final sedimentation (✓), primary sludge digestion (✓), storage of digested sludge (✓), dumping of digested sludge (✓).

Your score: ☐ out of 42

Your improvement index: $\dfrac{\boxed{}/42}{\boxed{}/61} \times 100\% = \boxed{}\%$

Index

Notes

Notes

Notes